HETEROCYCLES FROM TRANSITION METAL CATALYSIS

Catalysis by Metal Complexes

Volume 28

Editors:

Brian James, *University of British Columbia, Vancouver, Canada*
Piet W. N. M. van Leeuwen, *University of Amsterdam, The Netherlands*

Advisory Board:

The titles published in this series are listed at the end of this volume.

HETEROCYCLES FROM TRANSITION METAL CATALYSIS
FORMATION AND FUNCTIONALIZATION

by

ANDRÁS KOTSCHY

*Eötvös Loránd University,
Budapest, Hungary*

and

GÉZA TIMÁRI

*CHINOIN Co. Ltd. member of Sanofi-Aventis Group,
Budapest, Hungary*

 Springer

A C.I.P. Catalogue record for this book is available from the Library of Congress.

ISBN-10 1-4020-3624-8 (HB)
ISBN-10 1-4020-3692-2 (e-book)
ISBN-13 978-1-4020-3624-8 (HB)
ISBN-13 978-1-4020-3692-7 (e-book)

Published by Springer,
P.O. Box 17, 3300 AA Dordrecht, The Netherlands.

www.springeronline.com

Printed on acid-free paper

Printed in the Netherlands.

Dedication

*This book is dedicated to our wives
and children for their unrelenting
patience and support.*

Contents

Late transition metal catalysis in the functionalization of heterocycles.

Foreword

Three decades after their conception transition metal catalyzed carbon-carbon and carbon-heteroatom bond forming reaction became an indispensable tool for synthetic chemists including those whose interest is focused on heterocycles. Following the "application boom" of the ninties and the intense refining of the catalyst systems in the last decade name reactions such as the Suzuki coupling, Heck reaction, or Buchwald-Hartwig coupling became nearly as common as the Michael-addition or Friedel-Crafts acylation.

Extensive research in the field led to the publication of several reviews and monographs in the field. The place of our book in this niche is to provide an overview of the developments in the application of palladium, nickel and copper catalyzed transformations in the preparation and functionalization of heterocyclic compounds. Although preference was given to recent results, important examples of earlier (pioneering) works are also included in this monograph.

When releasing this book to the public, we would like to thank the colleagues, who helped in the preparation of the manuscript – András Nagy in particular, Árvácska Sárpátki, Beatrix Bostai, Júlia Dénes and Júlia Timári. András Kotschy also thanks the hospitality of Prof. Heinrich Wamhoff at the Kekülé Institute of Organic Chemistry, University of Bonn and the Alexander von Humboldt Foundation, while collecting material for the book.

Abbreviations

Δ	solvent heated under reflux
acac	acetylacetonyl
^{t}Am	2-methylbut-2-yl
9-BBN	9-borabicyclo[3.3.1]nonane
BINAP	2,2'-bis(diphenylphosphino)-1,1'-binaphthyl
bipy	2,2'-bipyridyl
Bn	benzyl
Boc	*tert*-butyloxycarbonyl
^{t}Bu	*tert*-butyl
Cbz	benzyloxycarbonyl
COD	1,4-cyclooctadiene
CuTC	copper thiophene-2-carboxylate
Cy	cyclohexyl
DBU	1,8-diazabicyclo[5.4.0]undec-7-ene
dba	dibenzylideneacetone
DCE	dichloroethane
DCM	dichloromethane
DDQ	2,3-dichloro-5,6-dicyano-1,4-benzoquinone
DIBAL	diisobutylaluminium hydride
DIPA	diisopropylamine
DMA	*N,N*-dimethylacetamide
DMAP	4-dimethylamino-pyridine
DME	1,2-dimethoxyethane
DMEDA	*N,N*'-dimethyl-ethylenediamine
DMF	*N,N*-dimethylformamide
DMSO	dimethylsulfoxide

dppb	1,4-bis(diphenylphosphino)butane
dppe	1,2-bis(diphenylphosphino)ethane
dppf	1,1'-bis(diphenylphosphino)ferrocene
dppp	1,3-bis(diphenylphosphino)propane
EDIPA	*N*-ethyl-diisopropylamine
HMPA	hexamethylphosphoramide
LDA	lithium diisopropylamide
LHDMS	lithium hexamethyldisilazane
LTMP	lithium 2,2,6,6-tetramethylpiperidine
NBS	*N*-bromosuccinimide
NMI	1-methylimidazole
NMP	1-methyl-2-pyrrolidinone
MW	microwave irradiation
Piv	2,2-dimethyl-propanoyl
SEM	2-(trimethylsilyl)ethoxymethyl
TBAF	tetrabutylammonium fluoride
TBDMS	*tert*-butyldimethylsilyl
TBS	*tert*-butyldimethylsilyl
Tf	trifluoromethanesulfonyl (triflyl)
TFA	trifluoroacetic acid
TFP	tri-*o*-furylphosphine
THF	tetrahydrofuran
TIPS	triisopropylsilyl
TMEDA	*N,N,N',N'*-tetramethyl-ethylenediamine
TMU	tetramethylurea
TMS	trimethylsilyl
Tol-BINAP	2,2'-bis(di-*p*-tolylphosphino)-1,1'-binaphthyl
Ts	4-methyl-phenylsulfonyl (tosyl)

Chapter 1

INTRODUCTION TO CATALYSIS

The second half of the twentieth century brought about a breakthrough in organic synthesis. Besides the constant improvement of classical procedures a wide array of unthought-of transformations were realized with the help of transition metals. Recognition of the importance of metal catalysis, especially transition metal catalysis, led to the opening of new avenues enabling the introduction of great molecular complexity, the improvement or in certain cases the complete reversal of the selectivity of classical reactions and the introduction of atom economical processes on the industrial level.

This book intends to give an overview of the development that was brought about by the acceptance and spreading of catalytic processes in synthetic heterocyclic chemistry. As transition metals constitute a major part of the periodic table, each of them having its own characteristic catalytic behaviour, we had to limit ourselves to a selection we coined "late transition metals" including palladium, copper and nickel. Even with these limitations the number of reactions to be discussed is too large to be included in this book, so we focus our attention on the characteristic features of catalysis in the formation and functionalization of heterocycles providing illustrative examples for the more important classes of heterocycles. For readers interested in the details of a certain process or the reactivity of a compound class there are a number of books[1] and reviews[2] available and references will also be made to relevant publications.

1.1 GENERAL CONSIDERATIONS

The application of transition metals in organic synthesis became an accepted and valued tool that broadened the scope of transformations

1

immensely. For someone entering the field for the first time the multitude of different transformations a transition metal (*e.g.* palladium) can facilitate through catalysis might seem frighteningly complex. A closer look into the details of the processes however reveals that their beauty lies in the simplicity, how key catalytic steps can be combined to result in seemingly complex transformations. With the understanding of the scope and limitations of the elementary steps one might design and execute state-of-the art reactions resulting in complex molecular structures.

The elementary steps common for most late transition metal catalyzed processes are listed in Table 1. These reactions, divided into three categories, form the basis of most of the transformations discussed. Any catalytic cycle (Figure 1-1.) should start with the attachment of a reactant to the catalyst (Entry). This is usually followed by the coordination or attachment of a second reagent to the catalyst (Attachment) providing proximity for transformations on the attached substrate or setting up the scene for the concluding step. There might be several "Attachment" steps in the process before reaching the concluding step (Detachment), where the product leaves the catalyst. If the catalyst is ready to enter this sequence after the Detachment step again, then we established a catalytic cycle, otherwise there has to be a follow up reaction where the active form of the catalyst is regenerated for the next Entry and the catalytic cycle is closed. Usually the form in which the transition metal is added to the reaction mixture is not active catalytically and in such cases the beginning of the catalytic cycle is preceded by the activation of the catalyst.

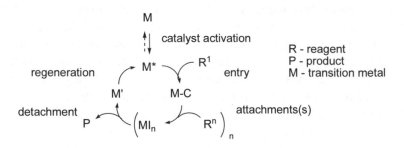

Figure 1-1. The general scheme of a catalytic cycle

The following chapters will discuss the characteristics of the elementary steps starting with those that form the Entry of catalytic cycles and finishing with those that appear usually in the Detachment step. It should be noted already here that most of the elementary steps functioning as Entry might also appear in the Attachment step (see Table 1.), although in general, combinations of alike steps are less common than other sequences.

Table 1. The key steps in catalytic processes

> **Entry**
> 1) Oxidative addition
> 2) Coordination-addition
>   *a*, Metal – nucleophile addition
>   *b*, Metal – hydrogen addition (hydrometalation)
> 3) C-H activation
> 4) Transmetalation
>
> **Attachment**
> 1) Transmetalation
> 2) Insertion
>   *a*, CO insertion
>   *b*, Olefin insertion (carbometalation)
> 3) C-H activation
> 4) Oxidative addition
>
> **Detachment**
> 1) Reductive elimination
> 2) *β*-hydride elimination

Of the steps listed in Table 1. some are encountered more frequently, while others are less common. Transition metal catalyzed processes usually begin with *oxidative addition* or *coordination-addition* as an Entry, which is commonly followed by *transmetalation* or *insertion* in the Attachment phase. The final Detachment step is either *reductive elimination*, or *β-hydride elimination,* depending on the nature of the intermediate.

1.2   THE ELEMENTARY STEPS OF CATALYTIC REACTIONS

This chapter discusses the most important characteristics of the elementary steps appearing in this book according to our present knowledge. These reactions, although we call them elementary steps are not to be confused with elementary reactions, as they might consist of more than one elementary reaction but are treated as a whole. We would like to call the attention of the reader to the fact that some of the rules of catalysis laid down in the past decades and accepted widely as a working model are based on empirical data, and although they are sufficiently detailed for a synthetic chemist, might only have limited validity.

As several elementary steps (e.g. transmetalation, C-H activation) might appear at different stages of the process they will be discussed in general, irrespective whether they open a catalytic cycle or appear only before it's conclusion.

Oxidative addition

By far the most common way for organic molecules to enter late transition metal catalyzed reactions is oxidative addition. In this process a low valent palladium(0)[3] or nickel(0) atom inserts into a carbon-heteroatom bond, usually of an aryl halide or sulfonate (Figure 1-2). The formation of the carbon-metal bond is accompanied by an increase in the oxidation number of the metal by 2. There are a series of factors determining the speed of the process.

$$R–X + L_nM^{(n)} \longrightarrow R–M^{(n+2)}–X$$
$$L_n$$

R: alkyl, aryl, vinyl, alkynyl X: I, OTf ($-OSO_2CF_3$), Br, Cl, OTs

$M^{(n)} \longrightarrow M^{(n+2)}$: Ni(0) $\longrightarrow$ Ni(2); Pd(0) $\longrightarrow$ Pd(2); Cu(1) $\longrightarrow$ Cu(3);

Figure 1-2. The general scheme of oxidative addition

The electron density of the metal atom has a profound effect on its ability to undergo oxidative addition. The more electron rich the metal centre, the faster the oxidative addition, which explains why nickel(0) complexes are in general more reactive than the analogous palladium(0) complexes. One way to increase the electron density on the metal is by the use of good donor ligands. Commonly used ligands that give highly reactive complexes include P^{t}Bu$_3$, PCy$_3$, 2-dialkylphosphino-biphenyls, bidentate ferrocenylphosphines and nucleophilic heterocyclic carbenes (Figure 1-3) formed *in situ* from imidazolium salts. In certain cases the catalytic activity of the ligand originating in its electron donating ability and steric bulk is further enhanced by its ability to form cyclopalladated complexes (*e.g.* tri-*o*-tolylphosphine).[4]

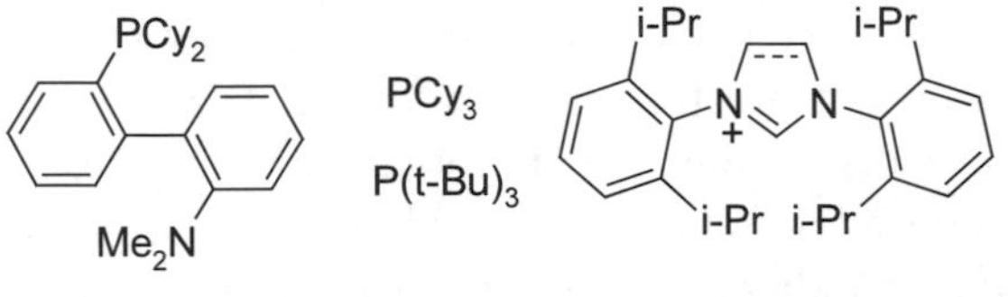

Figure 1-3. Some ligands that give highly active catalytic systems with palladium

The active form of the catalyst is coordinatively unsaturated, so its catalytic activity also depends on the ease of the formation of intermediates containing a low coordination metal. In general the more sterically demanding ligands are present, the more facile is the formation of the active catalyst. For example the outstanding activity of the Pd-P^{t}Bu$_3$ system was attributed to the fact that the active catalyst is a 1:1 complex[5] due to the increased steric demand of the ligand. *N.B.* PCy$_3$, which has a cone angle of "only" 170° (*c.f.* 182° for P^{t}Bu$_3$), has more room for the formation of the two extra bonds and gives an active catalyst already at a 1:2 Pd:P ratio.

The nature of the carbon-heteroatom bond into which the metal inserts is also influential. The most commonly used substrates are halides or triflates with a general reactivity order of I>OTf>Br>>Cl. The transformations of other sulfonates[6], fluorides[7] or anilinum salts[8] are also emerging as new alternatives but they usually require the use of the more active nickel catalysts. The electron density of the organic moiety also influences the ease of the oxidative addition. Aryl halides bearing electron withdrawing substituents are usually more reactive than their electron rich counterparts.

Finally, the hybridization of the carbon atom also has a marked effect on its willingness to attach to the transition metal. Allyl or benzyl halides undergo oxidative addition faster than aromatic or vinyl halides. The least reactive are alkyl halides which require the use of nickel(0)[9] complexes or highly active catalyst systems.[10] If we start from an optically active substrate, then the oxidative addition usually proceeds in a stereoselective manner.

Coordination - addition

In a series of late transition metal catalyzed processes the first step in the catalytic cycle is the coordination of the reagent to the metal atom, which is in a positive oxidation state, followed by its covalent attachment through the concomitant breaking of an unsaturated carbon-carbon bond or a carbon-hydrogen bond. These processes usually require a highly electrophilic metal centre and are frequently carried out in an intramolecular fashion. The carbometalation processes that follow a similar course, but take place only at a later stage in the catalytic cycle, will be discussed later.

Metal – nucleophile addition

Coordination of a carbon-carbon double bond to an electrophilic palladium centre makes the double bond sensitive towards the attack of nucleophiles. The nucleophile attacks the carbon-carbon bond from the *anti*-side[11] and leads to the formal addition of the palladium and the nucleophile onto the double bond (which is facilitated by the loosening of this bond through d-π* backdonation form palladium), the palladium being attached to the sterically less hindered site (Figure 1-4). The so formed palladium-alkyl complex might participate in a series of synthetically useful transformations including attachment reactions (transmetalation, insertion) or regeneration of the carbon-carbon double bond through β-hydride elimination.

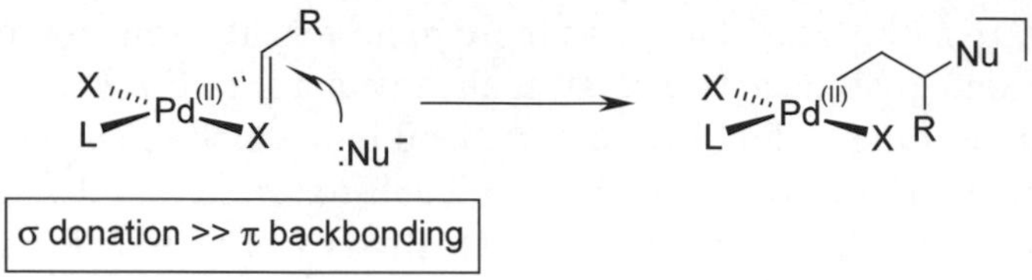

σ donation >> π backbonding

Figure 1-4. Palladium(II) assisted nucleophilic addition onto double bond

An illustrative example of the use of this process in the preparation of nitrogen heterocycles is presented in (**1.1.**) Stahl and co-workers reported[12] the synthesis of a series of pirrolidine derivatives exploiting the fact that δ-ethynyl-amides undergo ring closure in the presence of a palladium(II) catalyst, base and oxidant.

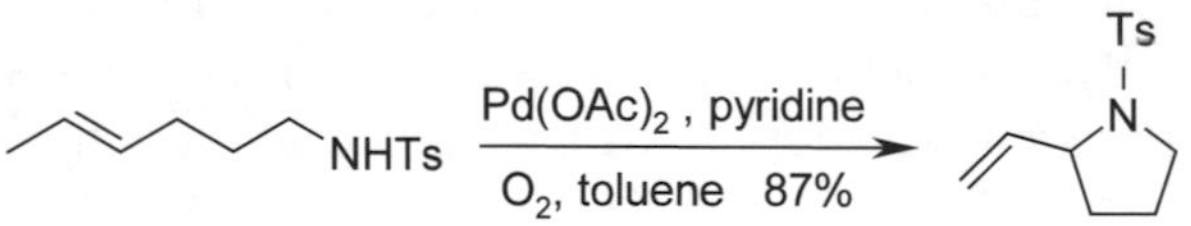

$$\text{(1.1.)}$$

Metal–hydrogen addition (hydrometalation)

An alternate approach for the attachment of an unsaturated reagent to a transition metal is the addition of a metal-hydrogen bond onto the carbon-carbon multiple bond. This equilibrium process (denoted hydrometalation) leads to an alkyl- or alkenylmetal complex (Figure 1-5) that might undergo a series of subsequent transformations. From the practical point of view the placement of the equilibrium is less interesting as long as it is able to provide enough of the alkyl- or alkenylmetal complex for the follow up reaction. The addition process is stereoselective and gives always the *syn*-adduct.

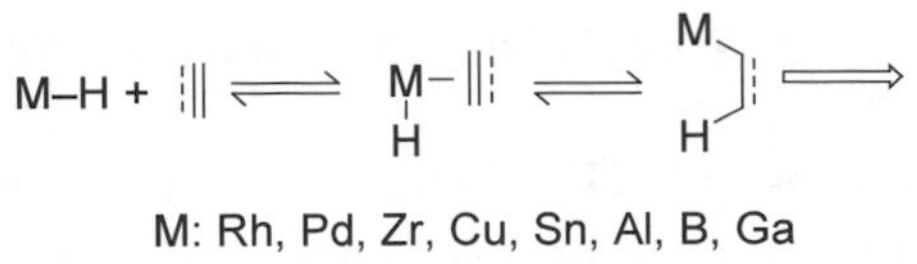

M: Rh, Pd, Zr, Cu, Sn, Al, B, Ga

Figure 1-5. The hydropalladation of multiple bonds

Triple bonds are in general more reactive than double bonds as is exemplified in the following process (**1.2.**).[13] The active catalyst is HPdOAc, which is formed by the oxidative addition of acetic acid onto Pd(0). The organic substrate is attached to the palladium in a regio- and stereospecific step that is followed by an oxidative addition (*N.B.* Pd(II)-Pd(IV) transition) and reductive elimination, or alternatively carbopalladation and reductive elimination, to give the indole derivative.

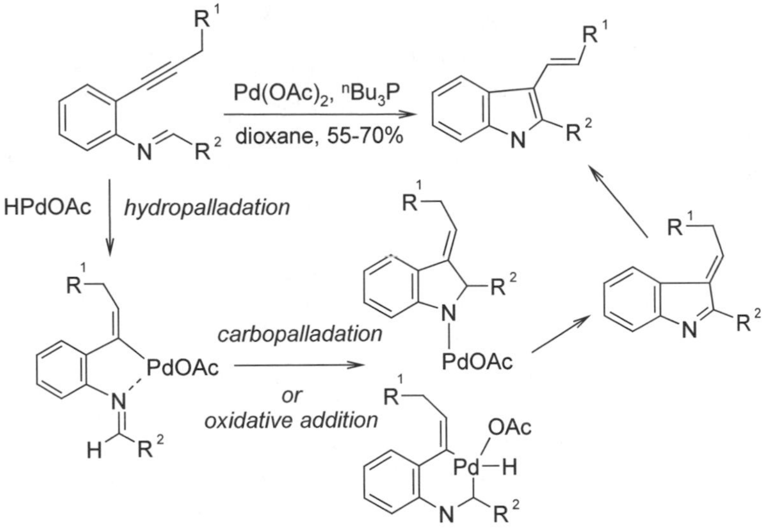

(1.2.)

C-H activation

The entries into transition metal catalysis discussed so far, required the presence of a specific bond (a polar carbon-heteroatom bond for oxidative addition or a carbon-carbon multiple bond for coordination-addition processes) that was sacrificed during the process. If we were able to use selected carbon-hydrogen bonds as sacrificial bonds, then we could not only save a lot of trouble in the preparation of starting materials but we would also provide environmentally benign alternatives to several existing processes. In spite of the progress made in this field the number of such transformations is still scarce compared to the aforementioned reactions.

There are however certain cases where the selective activation of a carbon-hydrogen bond is viable.

The key to the success of C-H activation is the fixing of the palladium(II) centre in the proximity of a carbon-hydrogen bond (usually sp^2 hybridized). This might take place through covalent bonding (intramolecular process) or with the help of non-covalent interactions. An example of each process is presented below.

In **1.3.** palladium is inserted into the carbon-iodine bond (oxidative addition), and in the resulting complex the palladium atom is held in the proximity of the carbon-hydrogen bond of the neighbouring ring (*n.b.* there is only one available C-H bond on the other ring). Following C-H activation palladium eliminates from the resulting complex to give the desired product and re-enters the catalytic cycle. [14]

(1.3.)

In the opening step of the catalytic cycle of the intermolecular variant[15] (**1.4.**) palladium is coordinated by the lone pair of the aniline nitrogen atom and is held in the proximity of the *ortho*-hydrogens (*c.f. ortho*-lithiation, Murai reaction). This results in the breaking of the carbon-hydrogen bond and the formation of a palladium-carbon bond and a molecule of acetic acid. This reaction is usually described either as an oxidative addition – reductive elimination sequence, or an electrophilic substitution. The formed arylpalladium complex than undergoes carbopalladation of the olefin and β-hydride elimination to give the product analogous to a Heck type process. In this reaction the palladium catalyst leaves the stage in the 0 oxidation state but enters it in the +2 oxidation state, so to close the catalytic cycle it has to be reoxidized.

(1.4.)

Transmetalation

Probably the most common attachment reaction in a late transition metal catalyzed reaction is transmetalation. This reaction, depicted in Figure 1-6, is the reversible exchange of covalently bonded ligands between two metal centres. The placement of the equilibrium is usually determined by the difference between the thermodynamic stability of the sacrificed and the formed bonds. From the practical point of view the placement of the equilibrium is less interesting as long as it is able to provide enough of the transmetalated complex for the follow up reaction.

$$M-X + M'-R \rightleftharpoons M-R + M'-X$$

$$Li \longrightarrow Mg, Al \longrightarrow Zn, Sn, Si, B \longrightarrow Ni, Cu \longrightarrow \underline{Pd}$$

Figure 1-6. The general transmetalation order of metals

In spite of its abundant use there is very little known about the actual mechanism of transmetalation. In certain cases the transmetalating agent has to be activated (made more nucleophilic) to achieve the desired conversion. Typical examples are boron- and siliciumorganic compounds, where the presence of hydroxide or fluoride ions is often necessary for a successful transformation.

Because of their frequent use, some late transition metal catalyzed carbon-carbon bond forming reactions evolved into name reactions. The most prominent examples are cross-coupling reactions, where distinction is usually made on the basis of the transmetalating agent used. The common mechanism of cross-coupling reactions and its "name variants" are discussed in Chapter 2.1.

Coordination-insertion

Late transition metals have a marked affinity towards coordinating carbon monoxide and carbon-carbon multiple bonds. If there is another suitable ligand on the metal centre, this coordination might be followed by the insertion of a carbonyl group or the carbon-carbon moiety into the metal-ligand bond. Both types of attachment reactions are commonly exploited in catalytic processes and their characteristics will be discussed separately.

Carbonyl insertion

In the process of carbonyl insertion the 1,1 migratory insertion of the coordinated CO ligand into the metal-carbon bond results in the formation of a metal-acyl complex (Figure 1-7). This process, as nearly all elementary steps discussed so far, is reversible, but even when using atmospheric CO pressure the equilibrium is mostly shifted towards insertion. In the process of insertion a vacant coordination site is also produced on the metal, where further reagents might be attached. Of the metals covered in this book palladium is by far the most frequently utilized in such transformations.

$$R{-}M + CO \xrightarrow{\text{coordination}} R{-}M{\cdots}CO \xrightarrow{\text{insertion}} \underset{R}{\overset{O}{\underset{\qquad}{\overset{\|}{C}}}}{\diagdown}M$$

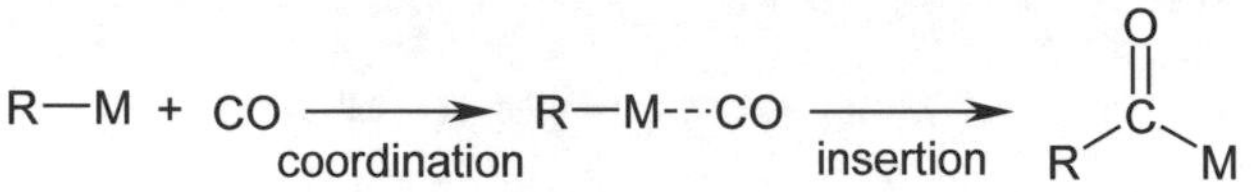

Figure 1-7 The general process of carbonyl insertion

An acyl-palladium complex might undergo a series of follow up reactions. Subsequent transmetalation and reductive elimination lead to the formation of a carbonyl compound. This process is also coined carbonylative coupling, referring to the cross-coupling reaction, which would take place in the absence of carbon monoxide under similar conditions (for more details see Chapter 2.4.).

If the reaction mixture also contains a nucleophile, then the acyl-palladium complex might undergo displacement of the metal, which usually leads to the formation of a carboxylic acid derivative. The side product in this process is a palladium(II) complex that undergoes reductive elimination to regenerate the catalytically active palladium(0) complex.

Olefin insertion (carbometalation)

In the process of olefin insertion, also known as carbometalation, the 1,2 migratory insertion of the coordinated carbon-carbon multiple bond into the metal-carbon bond results in the formation of a metal-alkyl or metal-alkenyl complex. The reaction, in which the bond order of the inserted C-C bond is decreased by one unit, proceeds stereoselectively (*syn*-addition) and usually also regioselectively (the more bulky metal is preferentially attached to the less substituted carbon atom. The willingness of alkenes and alkynes to undergo carbometalation is usually in correlation with the ease of their coordination to the metal centre. In the process of insertion a vacant coordination site is also produced on the metal, where further reagents might be attached. Of the metals covered in this book palladium is by far the most frequently utilized in such transformations.

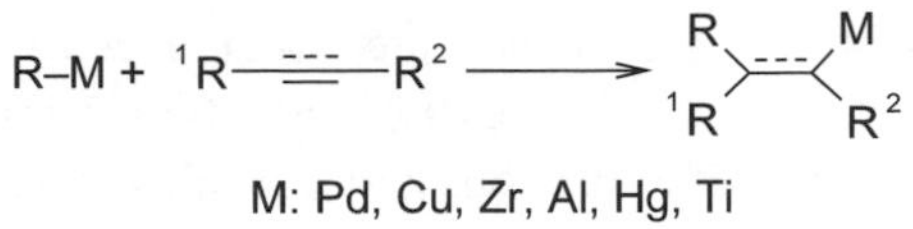

$$R-M + {}^1R \stackrel{---}{=\!=\!=} R^2 \longrightarrow$$

M: Pd, Cu, Zr, Al, Hg, Ti

Figure 1-8. The general equation of carbometalation

Intramolecular variants of this reaction are often utilized in the synthesis of (poly)cyclic systems, while the intermolecular variant of the transformation is the key step in one of the most frequently studied and utilized carbon-carbon bond forming reactions, the Heck reaction (for details see Chapter 2.2.).

In a catalytic cycle the carbometalation step is usually followed by β-hydride elimination. Other regular subsequent steps include carbometalation and transmetalation.

Reductive elimination

Probably the most common detachment step in late transition metal catalyzed processes is reductive elimination. In this transformation two groups, that are both attached to the same metal centre, will be released and form a covalent bond, with the concomitant formation of a metal whose formal oxidation state, coordination number and electron count are decreased by two. Figure 1-9 presents a general order of the ease of reductive elimination for the most common complexes.

When reductive elimination from a late transition metal involves the formation of a carbon-carbon bond, the process is intramolecular and the groups have to be aligned *cis* to one another in the complex. In the formation of carbon-heteroatom bonds the reductive elimination from palladium might take place via competing pathways.[16]

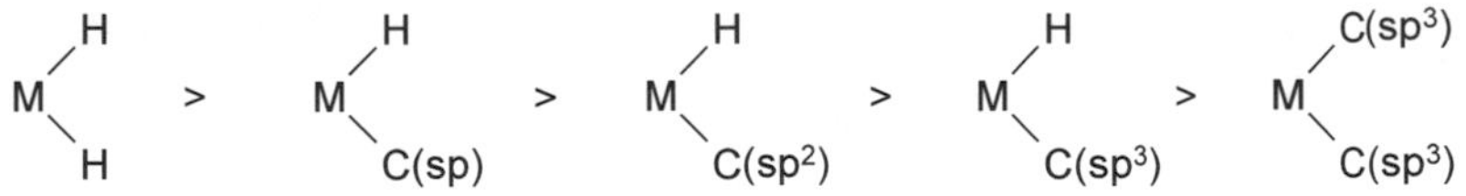

Figure 1-9. The general trend for reductive elimination from d^8 square planar complexes

The key factors that influence the ease of the reductive elimination by a late transition metal complex are listed in Figure 1-10. The understanding of their effect on the metal centre might help to design more active catalyst systems. The use of bulky ligands, for example might increase the crowdedness around the metal centre and facilitate reductive elimination, where the coordination number of the metal decreases by two units.[17] This is particularly true in reactions, where the reductive elimination is the rate determining step. The palladium catalyzed coupling of alkyl groups, or the formation of carbon-heteroatom bonds usually requires the use of a sterically demanding ligand (*n.b.* the increased cone angle of the ligand might also accelerate the oxidative addition step as seen before).

Similarly, decreasing the electron density on the metal will enhance its willingness to undergo reductive elimination and become more electron rich. A typical demonstration of this tendency is the nickel catalyzed coupling of *sp³* hybridized carbon atoms. Here, if a nickel-phosphine catalyst is used, the catalytic cycle is very slow, due to disfavoured reductive elimination. The use of additives that deplete the electron density of the nickel atom through coordination (*e.g.* fluorostyrenes) leads to a marked increase of the turnover frequency.[18]

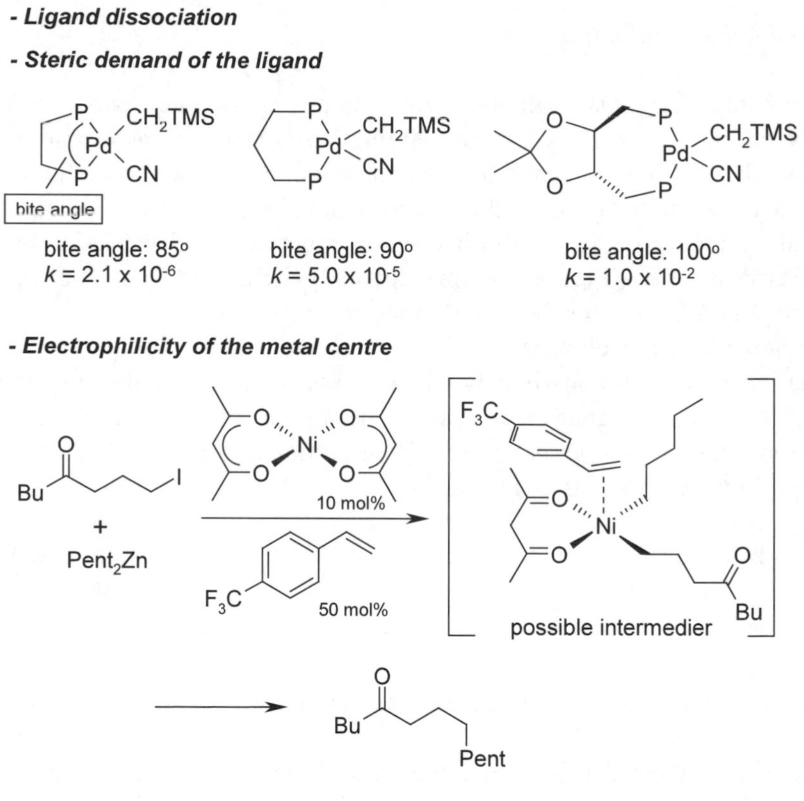

- Ligand dissociation

- Steric demand of the ligand

bite angle: 85°
$k = 2.1 \times 10^{-6}$

bite angle: 90°
$k = 5.0 \times 10^{-5}$

bite angle: 100°
$k = 1.0 \times 10^{-2}$

- Electrophilicity of the metal centre

Figure 1-10. Key factors determining the ease of reductive elimination

Kinetic studies also demonstrated, that the amount of coordinating ligands in the reaction mixture has a pronounced (reciprocal) influence on the rate of reductive elimination, therefore the metal-to-ligand ratio in catalytic systems has to be handled carefully too.[19]

β-hydride elimination

An alternative to reductive elimination for the detachment of the organic substrate from the metal catalyst is *β*-hydride elimination. In this process, as already suggested by its name, the metal catalyst and a hydrogen residing on the carbon atom *β* to the metal is eliminated from the organic molecule with the concomitant formation of a double bond. The olefinic product, coordinated to the metal, is usually not reactive under the applied conditions and dissociates readily from the metal centre, while the formed metal hydride complex in most cases undergoes reductive elimination of the hydrogen and an anionic ligand to generate a low oxidation state metal.

In the course of *β*-elimination (Figure 1-11) the metal has to have a vacant site *cis* to the alkyl substituent, where the hydride can coordinate. The reaction goes through a near planar transition state, which means that in substrates, where such a conformation is disfavoured by the organic moiety, the *β*-hidride elimination might become far too slow for practical applications.

The nature of the metal centre might also influence the speed of the *β*-elimination. Certain metals form a so called agostic interaction with the *β*-C-H bond which leads to its weakening and at the same time prearranges the ligand for the *β*-elimination. A typical example is shown in Figure 1-11.

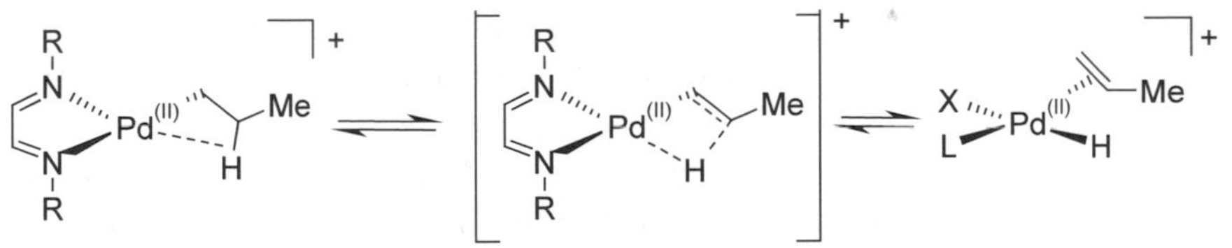

Figure 1-11. β-hydride elimination

From the practical point of view *β*-hydride elimination might also be an obstacle. In reactions that involve metal-alkyl complexes as early intermediates one has to block *β*-elimination to increase the lifetime of the intermediate and enable subsequent transformations on the complex. A reaction, which proved elusive partially for this very reason, is the coupling of alkyl halides. A set of conditions, which allowed for the Negishi coupling of primary alkyl halides and even tosylates with alkyzinc halides is shown in Equation **1.5**.[20] The recent work of Fu and others showed that the careful selection of the catalyst, additives and the reaction conditions might result in the facile formation of alkylpalladium and alkylnickel complexes that don't give β-elimination and open up the route for further reactions (*i.e.* transmetalation and reductive elimination).

$$R-X + R'-ZnBr \xrightarrow[\text{NMI, THF/NMP}]{\text{Pd}_2\,(\text{dba})_3,\,\text{PCy}_3} R-R'$$

$$(1.5.)$$

1.3   CLASSIFICATION OF THE CATALYTIC PROCESSES ON THE BASIS OF THE ELECTRONIC CHARACTER OF THE REAGENTS

For a synthetic organic chemist with little experience in catalysis the design of a process using the elementary steps of catalysis might be a bit too frightening, so we briefly outline another possible classification of the reactions discussed in this book, this time based on the nature of the reagents used (Table 2.). On the basis of the bond(s) that are "sacrificed" in the reaction one might divide the reagents into three (arbitrary) classes:

Reagents where a bond between a carbon atom of positive character and its electron withdrawing substituent is broken are classified as positive (+) and most frequently include a carbon-halogen, oxygen, nitrogen or sulphur bond.

Reagents where a carbon-carbon or a carbon-hydrogen bond is broken are classified as neutral (0) and most frequently include carbon-carbon multiple bonds or aromatic C-H bonds.

Reagents where a bond between a carbon atom of negative character and its low electronegativity substituent is broken are denoted as negative (-) and most frequently include carbon-metal, silicon or boron bonds.

The use of this classification might help to identify the ways a selected compound might participate in late transition metal catalyzed transformations, and might also help to establish potential reaction partners. Although Table 2 suggests that there is an abundance of potential reactions for a given substrate, we have to emphasize that certain classes are well represented in the synthetic literature (*e.g.* 3, 8, 11, 17), while for other classes there are only a very limited number of examples.

Table 2. Transition metal catalyzed reactions organized by the types of reagents involved. C_2 denotes the participation of an unsaturated carbon-carbon bond.

Class: + / +		*General equation:* **Ar-X + Ar'-X → Ar-Ar'**	
1)		(homo)coupling	*Catalyst:* Pd, Ni
2)		Radical (homo)coupling	*Catalyst:* Cu

Class: + / 0		*General equation:* **Ar-X + H-C → Ar-C**	
3)		Heck coupling	*Catalyst:* Pd
4)		Cyclopalladation (intramolecular)	*Catalyst:* Pd
5)	+CO	Carbonylative cyclopalladation	*Catalyst:* Pd
6)	+C$_2$	Carbopalladation-cyclopalladation	*Catalyst:* Pd
7)	+CO, C$_2$	Carbonylation,carbopalladation-cyclopalladation	*Catalyst:* Pd

Class: + / -		*General equation:* **R-X + Q-M → R-Q**	
8)	Q=R'	Cross-coupling	*Catalyst:* Pd, Ni
9)	+CO	Carbonylative coupling	*Catalyst:* Pd
10)	+C$_2$	Carbopalladation-coupling	*Catalyst:* Pd
11)	Q=NuH	Carbon-heteroatom bond formation	*Catalyst:* Pd, Ni
12)	+CO	Aminocarboxylation, alkoxycarboxylation	*Catalyst:* Pd
13)	+C$_2$	Carbopalladation C-N/O bond formation (ring anellation)	*Catalyst:* Pd
14)	+CO, C$_2$	"All in one" (ring anellation)	*Catalyst:* Pd, Ni

Class: 0 / 0		*General equation:* **Ar-H + Ar'-H → Ar-Ar'**	
15)		Oxidative coupling	*Catalyst:* Pd / 'O'
16)	+CO	Carbonylative oxidative coupling	*Catalyst:* Pd / 'O'

Class: 0 / -		*General equation:* **C-H + Q-M → C-Q**	
17)		Addition-elimination on alkenes ·	*Catalyst:* Pd / 'O'
18)	+CO	NuH Addition-carbonylation on alkenes	*Catalyst:* Pd / 'O'

Class: - / -		*General equation:* **R-M + Q-M → R-Q**	
19)		Carbon-heteroatom bond formation	*Catalyst:* Cu / 'O'

1.4   REFERENCES

[1] Recent books discussing late transition metal catalysis and its application in heterocyclic chemistry include:
(a) Li, J. J.; Gribble, G. W. *Palladium in Heterocyclic Chemistry: A Guide ot the Synthetic Chemist*, in *Tetrahedron Organic Chemistry Series*, Baldwin, J.; Williams, R. M. Eds., Pregamon, **2000**.
(b) Li, J. J. in *Alkaloids, Chemical and Biological Perspectives, Vol. 14*, Pelletier, S. W. Ed., Pergamon, Amsterdam, Netherlands, **1999**, 437-503.
(c) *Metal-catalyzed Cross-coupling Reactions*; Diedrich, F.; Stang, P. J. Eds. Wiley-VCH:

Weinheim, Germany, **1998**.

(d) Hegedus, L. S. *Transition Metals in the Synthesis of Complex Organic Molecules* 2nd Ed., University Science Books, Mill Valley, USA, **1999**.

(e) *Handbook of Organopalladium Chemistry for Organic Synthesis*, Negishi, Ei-ichi Ed. John Wiley & Sons, Inc., Hoboken (N.J), USA, **2002**.

(f) *Topics in Chemistry, Volume 219: Cross-coupling reactions*; Miyaura, N. Ed.; Springer, **2002**.

[2] Recent reviews discussing late transition metal catalysis in heterocyclic chemistry include:

(a) Stanforth, S. P. *Tetrahedron*, **1998**, *54*, 263.

(b) Beletskaya, I. P. *Pure and Applied Chemistry* **2002**, *74*, 1327.

(c) Dzhemilev, U. M.; Selimov, F. A.; Tolstikov, G. A. *ARKIVOC* **2001**, *2*, 85.

(d) Undheim, K.; Benneche, T. in *Adv. Heterocyclic Chem.* **1995**, *62*, 305.

(e) Kalinin, V. N. *Synthesis* **1992**, 413.

(f) Grigg, R.; Sridharan, V. *Pure and Applied Chemistry* **1998**, *70*, 1047.

(g) Godard, A.; Marsais, F.; Plé, N.; Trecourt, F.; Truck, A.; Quéguiner, G. *Heterocycles*, **1995**, *40*, 1055.

(h) Undheim, K.; Benneche, T. *Heterocycles* **1990**, *30*, 1155.

(i) Sakamoto, T.; Kondo, Y.; Yamanaka, H. *Heterocycles*, **1988**, *27*, 2225.

(j) *Chemical Reviews* **2004**, *104*, 2125-2812; a thematic issue dedicated to heterocycles.

(k) *Adv. Synth. Catal.* **2004**, *346*, 1503–1900, a thematic issue dedicated to cross-coupling reactions.

(l) Fairlamb, I. J. S. *Annu. Rep. Prog. Chem.,Sect. B* **2004**, *100*, 113.

[3] In certain cases palladium(II) complexes might undergo oxidative addition. For a review see: van der Boom, M. E.; Milstein, D. *Chem. Rev.* **2003**, *103*, 1759.

[4] Beller, M.; Fischer, H.; Herrmann, W. A.; Öfele, K.; Brossmer, C. *Angew. Chem. Int. Ed.* **1995**, *34*, 1848.

[5] Galardon, E.; Ramdeehul, S.; Brown, J. M.; Cowley, A.; Hii, K. K.; Jutand, A. *Angew. Chem. Int. Ed.* **2002**, *41*, 1760.

[6] (a) Kobayashi, Y.; William, A. D.; Mizojiri, R. *J. Organomet. Chem.* **2002**, *653*, 91. (b) Zim, D.; Lando, V. R.; Dupont, J.; Monteiro, A. L. *Org. Lett.* **2001**, *3*, 3049.

[7] Böhm, V. P. W.; Gsöttmayr, C. W. K.; Weskamp, T.; Herrmann, W. A. *Angew. Chem. Int. Ed.* **2001**, *40*, 3387.

[8] (a) Wenkert, E.; Han, A-L.; Jenny, C.-J. *J. Chem. Soc. Chem. Commun.* **1988**, 975. (b) Blakey, S. B.; MacMillan, D. W. C. *J. Am. Chem. Soc.* **2003**, *125*, 6046.

[9] Jensen, A. E.; Knochel, P. *J. Org. Chem.* **2002**, *67*, 79 and references therein.

[10] (a) Kirchhoff, J. H.; Dai, C.; Fu, G. C. *Angew. Chem. Int. Ed.* **2002**, *41*, 1945 and references therein. (b) Frisch, A. C.; Shaikh, N.; Zapf, A.; Beller, M. *Angew. Chem. Int. Ed.* **2002**, *41*, 4056.

[11] D. Milstein, J. K. Stille, *J. Am. Chem. Soc.*, **1978**, *100*, 3636.

[12] Fix, S. R.; Brice, J. L.; Stahl, S. S. *Angew. Chem. Int. Ed.* **2002**, *41*, 164.

[13] Takeda, A.; Kamijo, S.; Yamamoto, Y. *J. Am. Chem. Soc.* **2000**, *122*, 5662.

[14] Qabaja, G.; Jones, G. B. *J. Org. Chem.* **2000**, *65*, 7187.

[15] Boele, M. D. K.; van Strijdonck, G. P. F.; de Vries, A. H. M.; Kamer, P. C. J.; de Vries, J. G.; van Leeuwen, P. W. N. M. *J. Am. Chem. Soc.* **2002**, *124*, 1586.

[16] Driver, M. S.; Hartwig, J. F. *J. Am. Chem. Soc.* **1997**, *119*, 8323.

[17] Marcone, J. E.; Moloy, K. G. *J. Am. Chem. Soc.* **1998**, *120*, 8527.

[18] Giovannini, R.; Studemann, T.; Dussin, G.; Knochel, P. *Angew. Chem. Int. Ed.* **1998**, *37*, 2387.

[19] (a) Gillie, A.; Stille, J. K. *J. Am. Chem. Soc.* **1980**, *102*, 4933. (b) Gillie, A.; Stille, J. K. *J. Am. Chem. Soc.* **1981**, *103*, 2143.

[20] Zhou, J.; Fu, G. C. *J. Am. Chem. Soc.* **2003**, *125*, 12527.

Chapter 2

OVERVIEW OF CARBON-CARBON AND CARBON-HETEROATOM BOND FORMING REACTIONS

Although most transition metal catalyzed processes are built up of similar steps, they are usually divided into categories (sometimes name reactions) by the synthetic chemists. This classification is usually made on the basis of their synthetic utility rather than on mechanistic considerations. This chapter gives an overview of the most commonly used reactions, briefly outlining their mechanism as well as the scope and limitation of substrates in these processes.

2.1.   CROSS-COUPLING REACTIONS

Since their invention, mostly in the first half of the seventies, cross-coupling reactions had a spectacular career, and by now they are an accepted and appreciated tool of synthetic chemists.[1] They include a bunch of carbon-carbon bond forming reactions, usually between sp^2-sp^2 carbon centres, although recently some of them were extended to sp^3-sp^3 couplings too.[2] Figure 2-1 depicts the general mechanism of a cross-coupling reaction and lists the more common reagent combinations that became name reactions.

ML_n: Pd(0), Ni(0)   X: I, Br, Cl, OTf, OSO_2R, SOR, SR, $-N_2^+$,

M'			
B	Suzuki-Miyaura coupling	Sn	Stille coupling
Mg	Kumada-Karasch coupling	Zn	Negishi coupling
Cu	Sonogashira coupling	Si	Hiyama coupling

Figure 2.1. Cross-coupling reactions – mechanism and name reactions

The cross-coupling reactions begin with the oxidative addition of an aryl or vinyl halide or sulfonate onto a low oxidation state palladium or nickel atom, which is followed by the attachment of the other coupling partner to the metal in a transmetalation step. This is followed by an isomerisation step, which sets the stage for the final reductive climination. In this step the active, low oxidation state catalyst is regenerated and is ready to enter a next cycle. Cross-coupling reactions are usually distinguished on the basis of the transmetalating agent used. Common name reactions refer to different reagents: Kumada coupling – organomagnesium reagent, Negishi coupling – organozinc reagent, Stille coupling – organotin reagent, Suzuki coupling – organoboron reagent, Hiyama coupling – organosilicon reagent, Sonogashira coupling – organocopper reagent (usually limited to copper acetylides).

From the synthetic point of view, choice of the starting halide or sulfonate is usually based on availability. Reactivity and economy usually work antiparallel with bromides being the most frequently used substrates. The recent invention of highly active catalyst systems, on the other hand, broadened the applicability of aryl chlorides considerably in cross-coupling reactions.[3]

The choice of the transmetalating agent is usually determined by availability and functional group tolerance. These factors also work against one another, and good transmetalating agents (e.g. Grignard reagents, organozinc reagents) are less tolerant towards functional groups in general. Of the more functional group tolerant reagents organostannanes and organoboron reagents emerged recently. Their use, however usually requires the addition of a stochiometric amount of base or other additives. The use of organostannanes is also biased by their toxicity, including the side product formed in the coupling, although there are some promising developments in this direction too.[4]

2.2.   HECK REACTION

The Heck reaction,[5] sometimes also mentioned with cross-coupling reactions, deserves distinction not only for being mechanisticly different but also for its synthetic importance. In the catalytic cycle depicted in Figure 2.2. the organic moiety of an aryl or vinyl halide or sulfonate replaces a hydrogen attached to an olefinic sp^2-carbon atom.

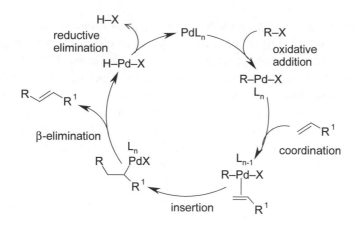

Figure 2-2. Mechanism of the Heck reaction

The first step in the cycle, analogous to the cross-coupling reactions, is the oxidative addition of an aryl (vinyl) halide or sulfonate onto the low oxidation state metal, usually palladium(0). The second step is the coordination of the olefin followed by its insertion into the palladium-carbon bond (carbopalladation). In most cases palladium is preferentially attached to the sterically less hindered end of the carbon-carbon double bond. The product is released from the palladium in a β-hydrogen elimination and the active form of the catalyst is regenerated by the loss of HX in a reductive elimination step. To facilitate the process an equivalent amount of base is usually added to the reaction mixture.

As the entry of the aryl halide into the catalytic cycle is analogous to the cross-coupling reactions, the scope and limitations in the choice of this reagent are alike in both cases. As far as the olefin is concerned, in general less substituted double bonds are carbometalated more easily. Typical procedures would employ an electron poor doble bond, such as an acrylate. In intramolecular processes the structure of the olefin poses less restrain on the success of the coupling. These reactions are frequently applied in the synthesis of polycyclic systems. An "extreme" example of the Heck reaction is given in 2.1. where the cyclopropane ring is formed in two subsequent carbometalation steps from an aryl iodide and allene.[6] As β-hydride elimination is not possible on the formed cyclopropylpalladium complex the concluding step of the process is the formation of a carbon-nitrogen bond.

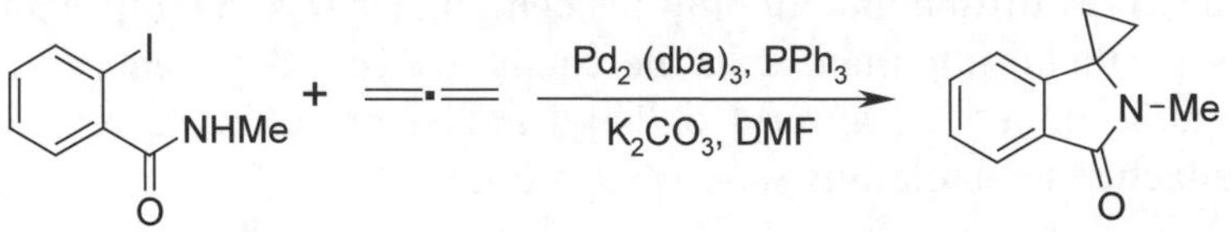

(2.1.)

Another variant of the Heck reaction which is important in heterocyclic chemistry utilizes five membered heterocycles as olefin equivalent (**2.2.**).[7] It is not clear whether the process, coined as "heteroaryl Heck reaction" follows the Heck mechanism (*i.e.* carbopalladation of the aromatic ring followed by β-elimination) or goes via a different route (*e.g.* electrophilic substitution by the palladium complex or oxidative addition into the C-H bond). Irrespective of these mechanistic uncertainties the reaction is of great synthetic value and is frequently used in the preparation of complex policyclic structures.

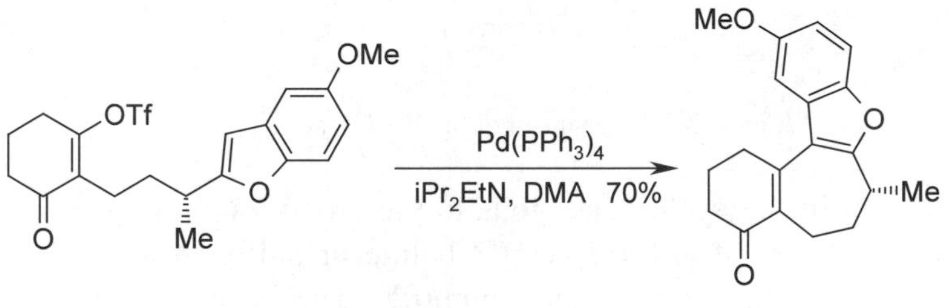

(2.2.)

2.3.   BUCHWALD-HARTWIG REACTION

In spite of their relatively young age palladium and nickel catalyzed carbon-heteroatom bond forming reactions, also known as the Buchwald-Hartwig reaction, have gained significant importance amongst synthetic heterocyclic chemists. The facile formation of carbon-nitrogen and carbon-oxygen bonds from aryl and vinyl halides, formally in a nucleophilic substitution, has opened up new, simple avenues and gained importance in fields as diverse as natural product synthesis, medicinal chemistry or materials science.

The seminal report of Migita[8] and its extension by Buchwald[9] revealed, that a cross-coupling reaction, in which the transmetalating agent is a metal amide is also feasible. Following this recognition Buchwald and Hartwig explored the scope and limitations of the palladium catalyzed formation of carbon-nitrogen and carbon-oxygen bonds.[10]

The opening step of the Buchwald-Hartwig reaction, similarly to the previous cases, is the oxidative addition of an aryl halide or sulfonate onto a low oxidation state metal. Although the term "Buchwald-Hartwig reaction" is usually reserved for palladium catalyzed processes, carbon-heteroatom bond formation also proceeds readily with nickel and copper. The nickel catalyzed processes follow a similar mechanism, while the distinctly different copper catalyzed reactions will be discussed in Chapter 2.5.

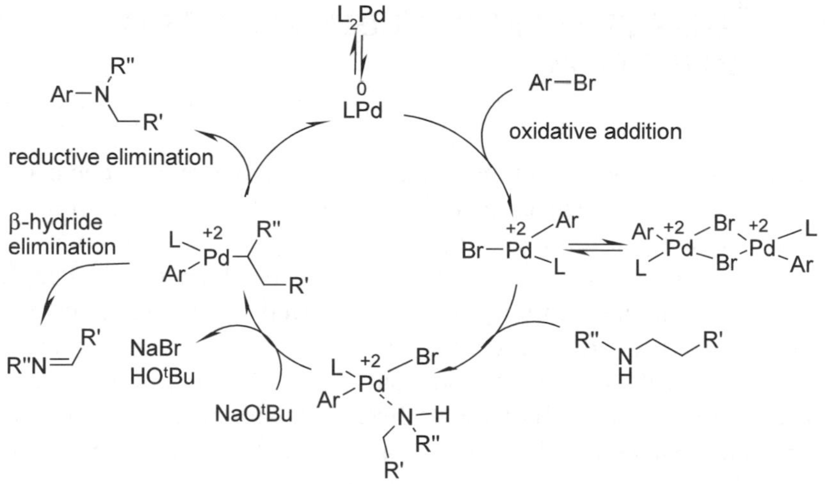

Figure 2-3. Mechanism of the Buchwald-Hartwig reaction

Following the oxidative addition, in the transmetalation step the anionic form of the heteroatom containing coupling partner (amide, alkoxide) is transferred onto the palladium, which is usually achieved by the combined use of the neutral form of the nucleophile and a suitable base. The choice of the proper base might be crucial for the success of the coupling. The transmetalation, as depicted in Figure 2-3, usually follows a coordination-attachment sequence (c.f. the simple attachment step in cross-coupling reactions).

The final step of the process is the detachment of the product from the metal in reductive elimination. Unlike in most cross-coupling reactions, this step was the limiting factor in the early reports on the Buchwald-Hartwig reaction. The use of sterically demanding mono and bidentate ligands, however helped to overcome this difficulty by facilitating the closing step.

As the entry of the aryl halide into the catalytic cycle is analogous to the cross-coupling reactions, the scope and limitations in the choice of this reagent are alike in both cases. The only difference worth mentioning is that in certain cases the added base might interfere with the use of perfluoroalkyl sulfonates. Carbon-nitrogen bond forming reactions are abundant and aliphatic amines can be introduced as well as aromatic amines and amides. The scope of carbon-oxygen bond forming reactions is more limited as the intermediate palladium alkoxyde complexes might undergo undesired side reactions (*c.f.* the palladium catalyzed oxidation of alcohols). Intramolecular variants of the Buchwald-Hartwig reaction usually proceed readily too.

2.4.   REACTIONS PROCEEDING WITH CO INSERTION

Carbon monoxide, a common ligand in organometallic chemistry, is known to insert into palladium-carbon bonds readily. This feature of the metal is frequently utilized when palladium catalyzed reactions are run in the presence of CO. The products of such reactions, also known as carbonylative couplings, incorporate a carbonyl group between the coupling partners.

The catalytic process (Figure 2-4) usually begins with the oxidative addition of an aryl halide or sulfonate onto the active form of the catalyst. In the presence of carbon monoxide the formed palladium-carbon bond breaks up with the concomitant insertion of a CO unit to give an acylpalladium complex. Such complexes might also be formed by the oxidative addition of acyl halides onto palladium.

The fate of the acyl palladium complex depends on the circumstances. In the presence of a suitable nucleophile (alcohol, amine) it is converted into the corresponding carboxylic acid derivative. The side product, a palladium hydride is converted to the active form of the catalyst in a reductive elimination step, resulting in the formation of an equimolar amount of acid, which is quenched by an added base (in most cases the excess of the nucleophile).

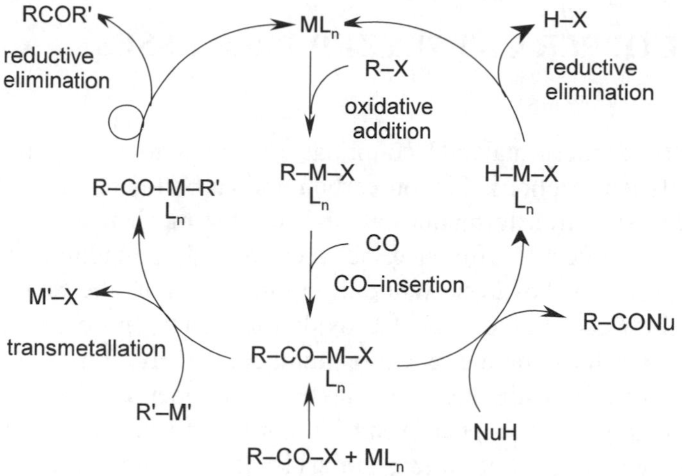

Figure 2-4. Carbonilative couplings

If the reaction mixture contains a reagent that is capable of transmetalation, then the other coupling partner is transferred onto the palladium this way. The formed complex, after ensuring the *cis* alignment of the two moieties in a quick isomerisation step, undergoes reductive elimination to give the desired carbonyl compound and regenerate the active form of the catalyst.

Carbonyl insertion can be coupled with a series of other transformations (e.g. C-C insertion, olefin functionalization) and leads to the introduction of significant structural complexity in a single step. An example of such a cascade is given in **2.3.** In this reaction the oxidative addition of the aryl halide is followed by the insertion of carbon monoxide. The acyl palladium complex undergoes intramolecular olefin insertion to give an alkyl palladium complex. In principle this complex could undergo β-hydride elimination (*c.f.* Heck reaction) but insertion of another CO molecule is faster and another acyl palladium complex is formed, which reacts with the added nucleophile to give the product. It is worth pointing out that the intramolecular insertion of the olefin prevails over the nucleophilic quenching of the first acyl palladium complex.[11]

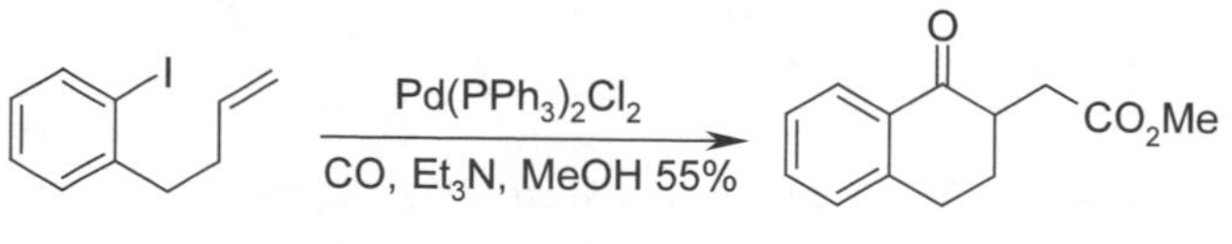

(2.3.)

2.5.   COPPER CATALYZED PROCESSES

Formally copper catalyzed couplings are analogous to palladium and nickel catalyzed reactions. Carbon-carbon and carbon-heteroatom bonds can be formed in such transformations alike. From the mechanistic point of view there is a significant difference between nickel, palladium and copper catalyzed processes however. While in the former cases the catalyst usually oscillates between the 0 and +2 oxidation states, in copper mediated transformations the common oxidation numbers are +1, +2 and +3.

There are two distinguishable classes of copper catalyzed processes. Reactions catalyzed by copper(I) salts (Figure 2-5) usually couple a reagent capable of oxidative addition (*e.g.* an aryl halide) and another reagent that can attach itself to the copper *via* transmetalation/coordination. Common examples include nitrogen, oxygen and carbon nucleophiles, such as alcohols, amines, azoles, cyanide and malonates. In these processes the sequential attachment of the coupling partners to the copper centre results in the formation of a copper(III) complex, which, on release of the product in reductive elimination, returns to the catalytically active +1 oxidation state. It is not always clear if the oxidative addition or the transmetalation step is the opening of the catalytic cycle as most nucleophiles exhibit a strong affinity towards copper salts and form stable complexes.

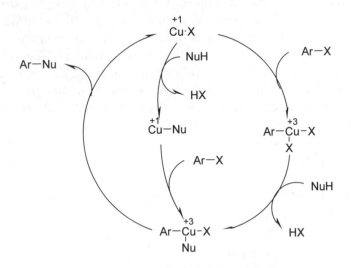

Figure 2-5. Copper(I) catalyzed coupling reactions

The other class of copper catalyzed reactions, utilizing copper(II) salts follows a different route. A typical example is the nucleophilic displacement reaction of arylboronic acids (Figure 2-6). Here both reagents are attached to the copper in transmetalation which leads to a copper(II) centre. This, however does not undergo reductive elimination, so these reactions have to be run in the presence of a suitable oxidizing agent, usually molecular oxygen. Oxygen transforms copper(II) into the copper(III) state, where reductive elimination proceeds readily to result in the formation of the desired product and a copper(I) salt, which is reoxidized to copper(II), the active form of the catalyst.

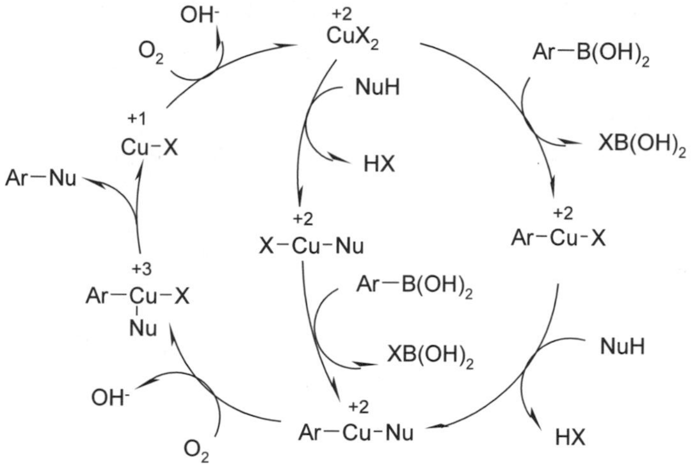

Figure 2-6. Copper(II) catalyzed coupling of arylboronic acids with nucleophiles

2.6. REFERENCES

[1] The cross-coupling reactions were reviewed recently by several authors including:
   (a) Hassan, J. Sévignon, M.; Gozzi, C.; Schulz, E.; Lemaire, M. *Chem. Rev.* **2002**, *102*, 1359.
   (b) Stanforth, S. P. *Tetrahedron*, **1998**, *54*, 263.
   (c) Negishi, E.; Anastasia, L. *Chem. Rev.* **2003**, *103*, 1979.
   (d) Sonogashira, J. *J. Organomet. Chem.* **2002**, *653*, 46.
   (e) Hiyama, T.; Shirakawa, E. *Top. Curr. Chem.* **2002**, *219*, 61.
   (f) Fairlamb, I. J. S. *Annu. Rep. Prog. Chem.,Sect. B* **2004**, *100*, 113.
[2] Zhou, J.; Fu, G. C. *J. Am. Chem. Soc.* **2004**, *126*, 1340 and references therein.

[3] Selected developmets include (a) phosphine oxide: Li, G. Y. *Angew. Chem. Int. Ed.* **2001**, *40*, 1513. (b) imidazolium salt: Navarro, O.; Kelly, R. A.; Nolan, S. P. *J. Am. Chem. Soc.* **2003**, *125*, 16194. (c) bulky phosphines: Yin, J.; Rainka, M. P.; Zhang, X.-X.; Buchwald, S. L. *J. Am. Chem. Soc.* **2002**, *124,* 1162.

[4] Fouquet, E.; Pereyre, M.; Rodriguez, A. L.; Roulet, T. *Bull. Chem. Soc. Fr.* **1997**, *134*, 959.

[5] Heck, R. F.; Nolley, J. P. Jr. *J. Org. Chem.* **1972**, *37*, 2320.

[6] Grigg, R.; Kordes, M. *Eur. J. Org. Chem.* **2001**, 707.

[7] Chambers, C. H.; Trauner, D. *Angew. Chem. Int. Ed.* **2002**, *41*, 1569.

[8] Kosugi, M.; Kameyama, M.; Sano, H.; Migita, T. *Chem. Lett.* **1983**, 927.

[9] Guram, A. S.; Buchwald, S. L. *J. Am. Chem. Soc.* **1994**, *116*, 7901.

[10] For leading reviews see (a) Hartwig, J. F. *Acc. Chem. Res.* **1998**, *31*, 852. (b) Wolfe, J. P.;Wagaw, S.; Marcoux, J. F.; Buchwald, S. L. *Acc. Chem. Res.* **1998**, *31*, 805. (d) Hartwig, J. F. *Angew. Chem. Int. Ed.* **1998**, *37*, 2046. (d) Kirsch, G.; Hesse, S.; Comel, A. *Current Organic Synthesis* **2004**, *1*, 47.

[11] Nieman, J. A.; Ennis, M. D. *J. Org. Chem.* **2001**, *66*, 2175.

Chapter 3

THE SYNTHESIS OF FIVE MEMBERED RINGS

The transition metal catalyzed synthesis of five membered heterocycles, particularly of condensed ring systems, has attracted considerable attention. The ease of the formation of five membered rings has been utilised both in intramolecular ring closure processes, and in the combination of two (three) fragments through the formation of a carbon-carbon and a carbon-heteroatom bond. This chapter is dedicated to examples, where the construction of the five membered heterocycle is achieved in a transition metal catalysed step.

3.1   TRANSMETALATION ROUTE

The formation of five membered heterocycles through cross-coupling reaction is a rare process. Although several procedures rely on the use of such reactions for the connection of two fragments through a carbon-carbon bond, the final ring closing step is usually the formation of a carbon-heteroatom bond through addition or substitution. These reactions will be discussed in Chapters 3.3. and 3.4. Examples of the transition metal catalysed formation of five membered heterocycles are focused around the preparation of condensed ring systems, and simple rings are still prepared mostly by classical synthetic methods.

An nice example of the formation of a five membered ring in cross-coupling reaction is the hexamethyldistannane mediated ring closure of the dihalogenated diaryl ether shown in **3.1**. The first step of the process is the palladium catalyzed exchange of one of the iodides to a trimethylstannyl moiety, followed by the closure of the five membered ring in Stille coupling.

The method enabled the efficient preparation of all four benzofuropyridine isomers.[1]

(3.1.)

The palladium catalyzed intramolecular coupling of aryl halides and classical carbanions, sometimes considered a variant of the Buchwald-Hartwig coupling, might also be used for the formation of heterocyclic systems. *N*-(2'-bromophenyl)-propionamides were converted in the presence of the appropriate palladium catalyst and lithium hexamethyldisilazide to oxindoles (**3.2.**). Under the applied conditions a series of electron deficient and electron rich aniline derivatives, including 2-chloroanilines were transformed successfully.[2]

(3.2.)

The examples presented in **3.3.-3.6.** describe the preparation of condensed ring systems containing five membered heterocycles, where the connection of the fragments is achieved in a similar manner. Formation of the carbon-carbon bonds by cross-coupling is accompanied by the formation of the carbon-nitrogen bond using classical organic transformations (Schiff-base formation, substitution etc.). Although the reactions where carbon-carbon bond formation precedes formation of the carbon-nitrogen bond formally fall outside the scope of this chapter, due to the synthetic importance of this approach some selected examples are mentioned here.

Indoles were prepared by Suzuki utilising the coupling of 2-iodoanilines and 2-ethoxyvinyl-boronates. The formed intermediates cyclized readily in the presence of acid, to give 3-substituted indoles in good yield (**3.3.**).[3] By "masking" the aniline as a 2-bromonitrobenzene derivative, Buchwald prepared a series of highly substituted indole derivatives. The coupling of these compounds with acetophenone, followed by the titanium mediated reduction of the nitro group resulted in the spontaneous closure of the indole ring (**3.4.**).[4]

N-protected *o*-anilineboronic acid was used in a series of experiments by Timári and co-workers to introduce an annelating indole moiety onto other heterocyclic systems. The tetracyclic indoloquinoline alkaloid Cryptosanguinolentine was prepared, for example, by the coupling of 3-bromoquinoline and an *N*-protected *o*-anilineboronic acid, followed by the conversion of the amino group to azide and ring closure of the thermally generated nitrene (**3.5.**).[5] The analogous reaction sequence starting from 2-quinolyl triflate leads to the formation of a nitrogen-nitrogen bond in the last step, providing access to indazolo[2,3-*a*]quinoline.[6]

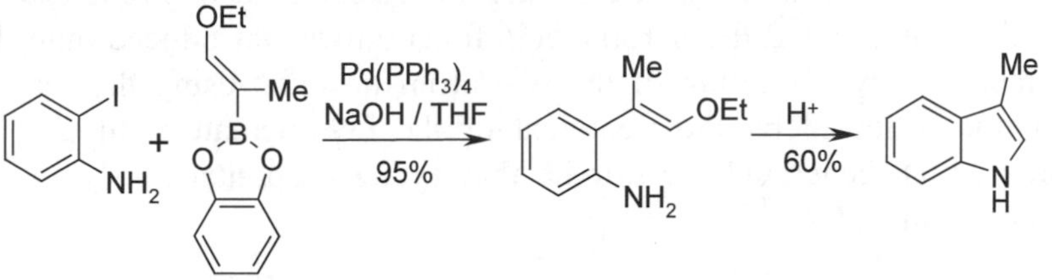

(3.3.)

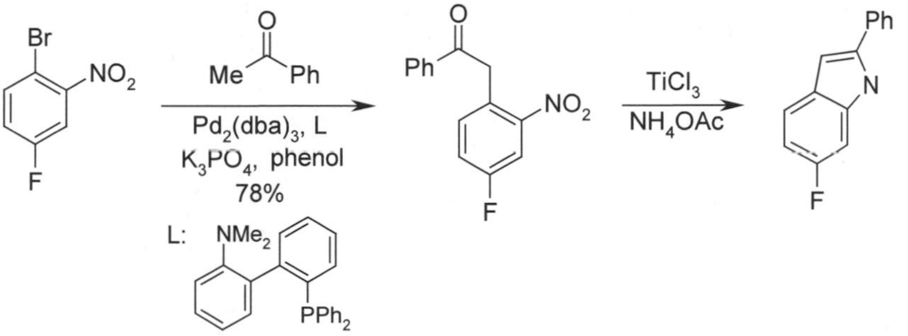

(3.4.)

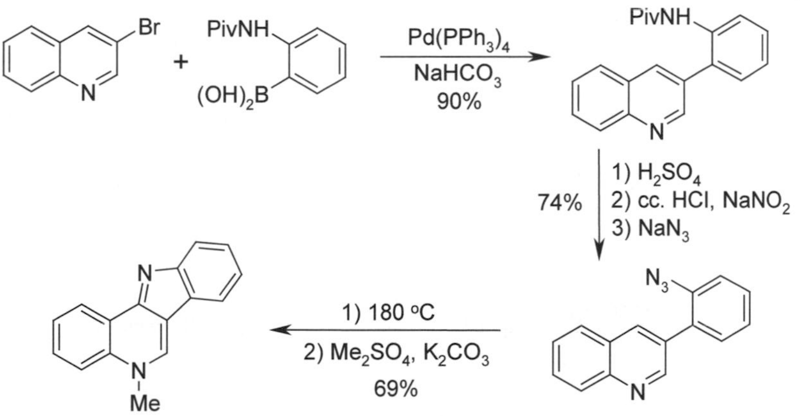

(3.5.)

Attempts to prepare the isomeric indoloisoquinoline system in the coupling of 3-isoquinolyl triflate and *o*-anilineboronic acid led to an unexpected result. In the course of the follow-up reactions the nitrogen atom in the isoquinoline ring had to be protected as the *N*-oxide to divert the cyclization of the nitrene into the 4-position. The intermediate diazonium salt, however under certain conditions underwent a spontaneous ring closure to the benzofuro[3,2-*c*]isoquinoline system, which on heating rearranges to the benzisoxazolo[2,3-*a*]isoquinoline (**3.6.**).[7]

In the preparation of the furocarbazole alkaloid, furostifoline, which followed a similar synthetic strategy, *o*-anilineboronic acid was replaced by 2-bromonitrobenzene and the boronic acid function was introduced onto the benzofurane moiety. Coupling of the two aromatic units using the Suzuki protocol was followed by the treatment of the intermediate with triethyl phosphite, which generated the nitrene that cyclized spontaneously to the tetracyclic product (**3.7.**).[8]

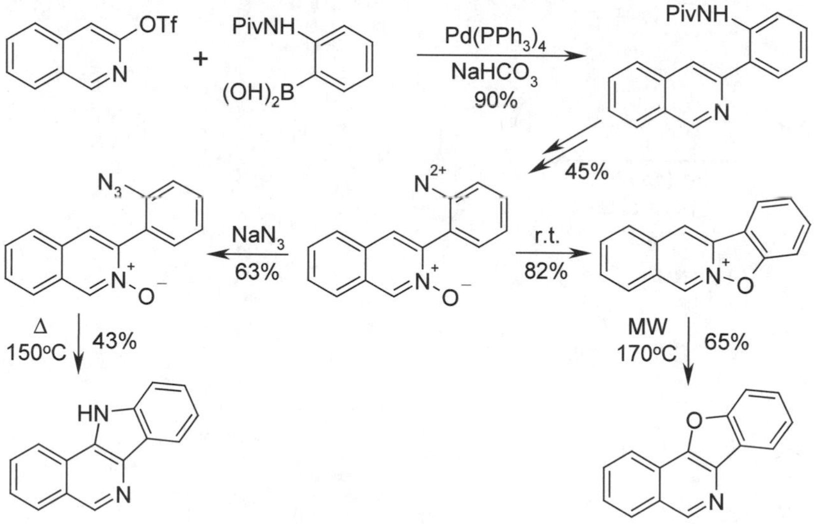

(3.6.)

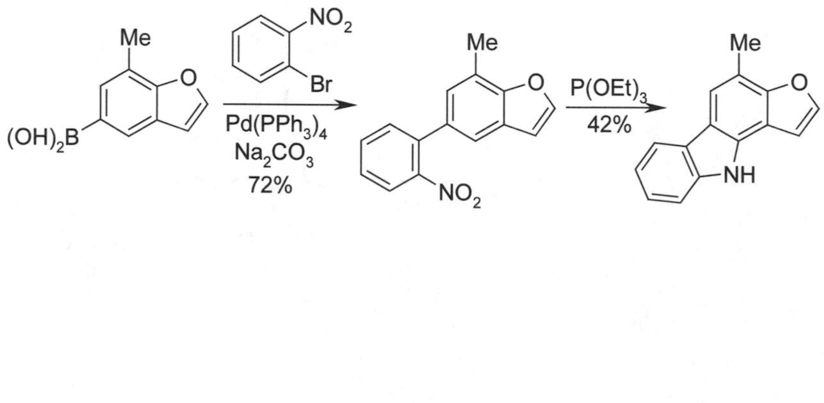

(3.7.)

3.2   INSERTION ROUTE

Transition metal catalyzed insertion reactions offer a convenient route for the preparation of five membered heterocyclic rings. Besides intramolecular Heck-couplings and CO insertion, examples of the intramolecular insertion of an acetylene derivative constitute the majority of this chapter. Although some of these processes involve the formation of a carbon-heteroatom bond, they are discussed here.

Heck reaction (olefin insertion)

Attachment of a pendant alkene moiety to the nitrogen atom of o-haloanilines and related compounds opens up the way for the palladium catalysed formation of a pyrrole ring. This approach has been utilised extensively in the preparation of indole derivatives. The following examples were divided by the type of the olefin side chain, where ring closure takes place. *N*-Allylanilines, such as *N*-allyl-2-iodoaniline undergo ring closure under a variety of conditions to give 3-methylindole in good to excellent yield (**3.8.**).[9] The reaction has also been extended successfully to functionalised allylanilines[10] and polycyclic analogues.[11]

The primary product of the Heck type ring closure is an indole derivative containing an exocyclic double bond. This compound, however in most cases isomerises spontaneously to the aromatic isomer. In the example shown in **3.9.** the addition of silver carbonate prevented the isomerisation of the double bond into the five membered ring.[12]

The palladium catalysed formation of indole derivatives has been extended by Grigg, who used carbon monoxide and unsaturated amines to trap the palladium complex formed in the insertion step. Reaction **3.10.** provides an example of such a transformation. The amides were converted to the cyclic derivatives using ring closing metathesis.[13]

$$\text{Pd(OAc)}_2$$
$$\text{Et}_3\text{N, MeCN}$$
$$87\%$$

(3.8.)

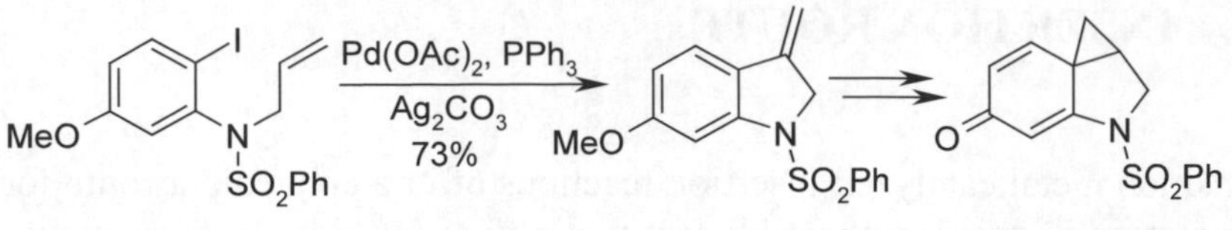

(3.9.)

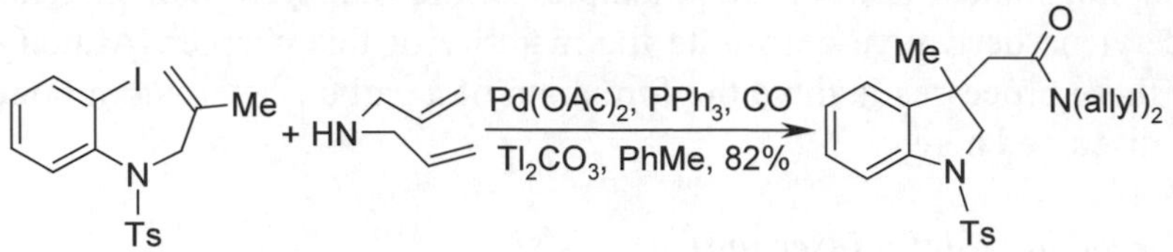

(3.10.)

Indolylacetic acid derivatives were prepared by the ring closure of the appropriate *N*-(γ-crotyl)-*o*-haloaniline derivative (**3.11.**).[14] Although the reported yields of the original transformations[15] were only mediocre, serious efforts were made to improve the method, due mainly to the value of the formed products in medicinal chemistry.

The same methodology was extended to the synthesis of other condensed pyrrole derivatives too. *N*-Boc 2-amino-3-iodothiophene was alkylated with ethyl 4-bromocrotylate to give the *N*-crotyl derivative, which on treatment with a palladium-triphenylphosphine catalyst cyclized efficiently to the appropriate pyrrolo[2,3-*b*]thiophene (**3.12.**).[16]

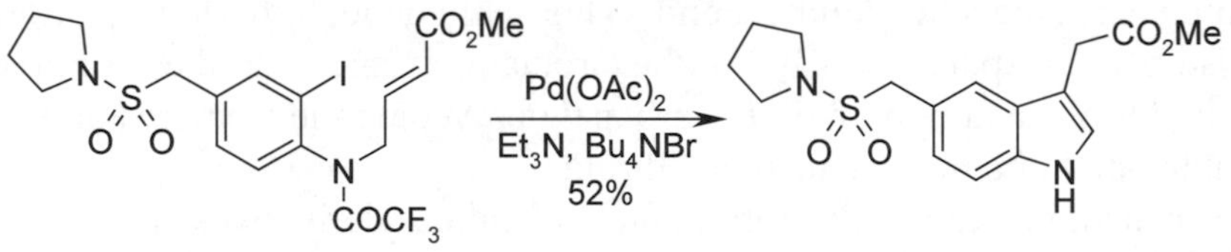

(3.11.)

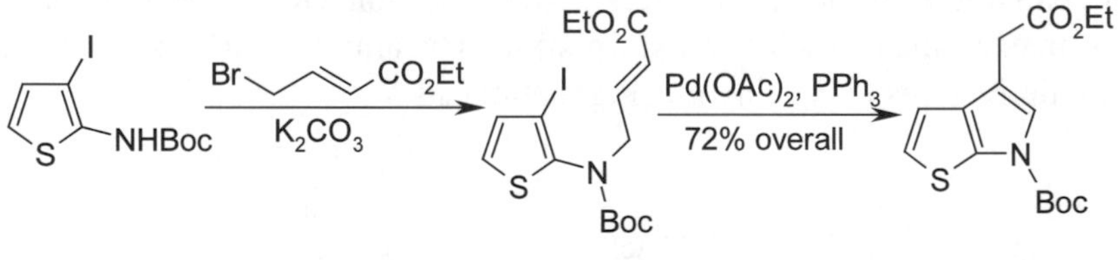

(3.12.)

The inclusion of the allyl moiety in an appropriate ring system led to the formation of a new, chiral quaternary carbon center (**3.13.**). By the use of optically active starting material, Mori and co-workers were able to control the stereochemistry of the new chiral centre (*N.B.* the chirality of the first centre was also set by a palladium catalysed reaction, the Tsuji-Trost allylation).[17]

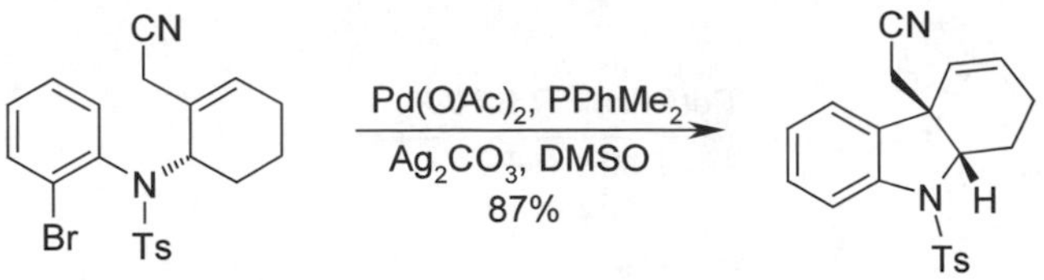

(3.13.)

Similar to *N*-allylanilines, their ease of formation makes *N*-acryloylanilines also an attractive starting material for the preparation of indole derivatives. Acrylates having an α-substituent give rise to chiral oxindole derivatives, both a common building block of natural products and a frequently employed synthon *en route* to them.[18] By using a chiral palladium-BINAP catalyst Overmann was able to achieve high enantioselectivity in the transformation shown in **3.14**.[19]

The same approach was applied in the enantioselective total synthesis of a complex natural product, quadrigemine C. The key step of the reaction sequence, establishing the chirality in the molecule is shown in **3.15**.[20] Besides the regular Heck coupling product a minor product was also identified in the process arising from the β-elimination of a methoxide moiety instead of the hydride.

The capture of the palladium complex formed in the intramolecular insertion of *N*-acryloyl-2-haloanilines has also been exploited by Grigg. The addition of aryl-, and vinylboron reagents to the mixture of *N*-(2'-methylacryloyl)-2-iodoaniline and a palladium catalyst led, after the closure of the five membered ring, to the transfer of the organic moiety to the formal terminal carbon atom of the acryloyl chain (**3.16.**). The role of the substituent in the 2'-position is presumably to block β-hydride elimination and so prolong the lifetime of the palladium complex formed in the insertion step.[21]

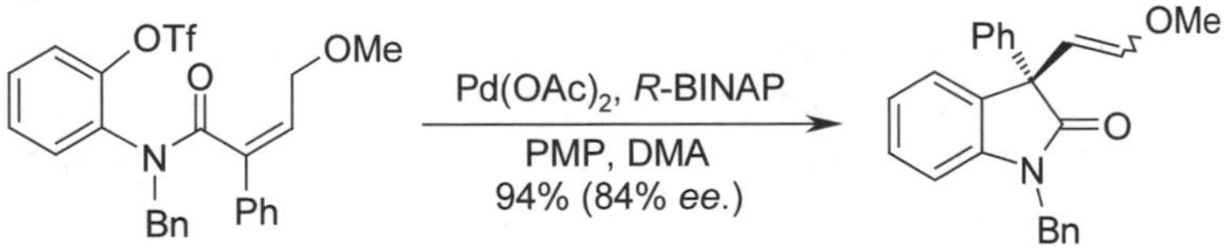

(3.14.)

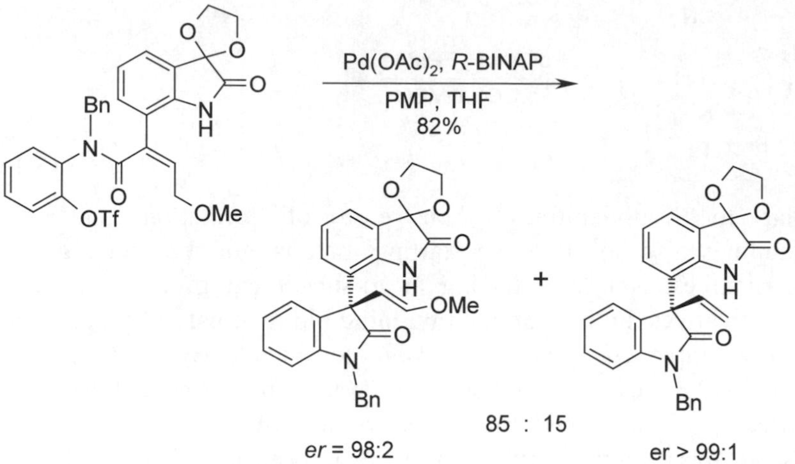

(3.15.)

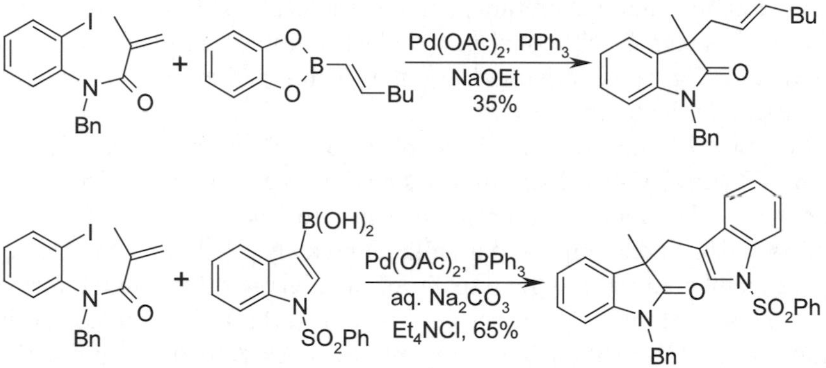

(3.16.)

Enamines, formed in the condensation of *o*-haloanilines and carbonyl compounds, might also undergo ring closure in the presence of a suitable palladium catalyst. 2-Iodoaniline was condensed with a series of cycloalkanone-2-carboxylic esters and the resulting 2-anilino-cycloalkene-1-carboxylic esters were cyclized in the presence of a palladium-triphenylphosphine catalyst and silver phosphate. This latter additive was crucial to obtain an effective coupling (**3.17.**). The yield of the reaction showed a marked dependence on the ring size, which might be attributed to the varying ease of the endocyclic double bond isomerisation that precedes the insertion.[22] The one-pot condensation and coupling of other 2-iodoaniline derivatives and cycloalkanones also showed a similar trend.[23]

The facile formation and ring closure of enamines has also been exploited in the solid-phase preparation of indolecarboxylates. Attachment of 2-iodoaniline or 2-bromoaniline to the resin in the form of an *α*-anilino-acrylate followed by the palladium catalysed formation of the five membered ring and concluded by the detachment of the formed indolecarboxylate from the resin led to the formation of a set of similar compounds (**3.18.**).[24]

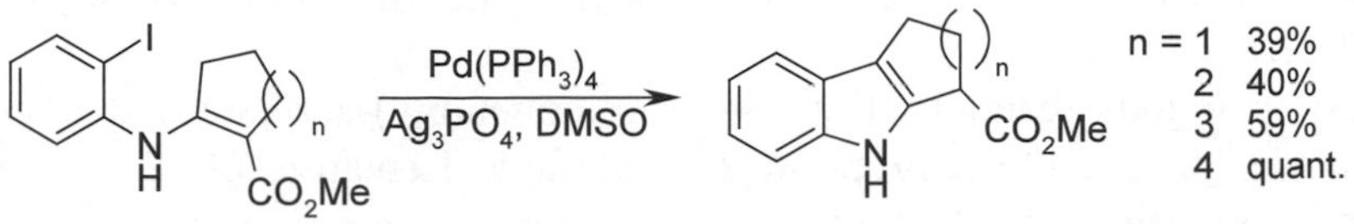

$$n = 1 \quad 39\%$$
$$2 \quad 40\%$$
$$3 \quad 59\%$$
$$4 \quad \text{quant.}$$

(3.17.)

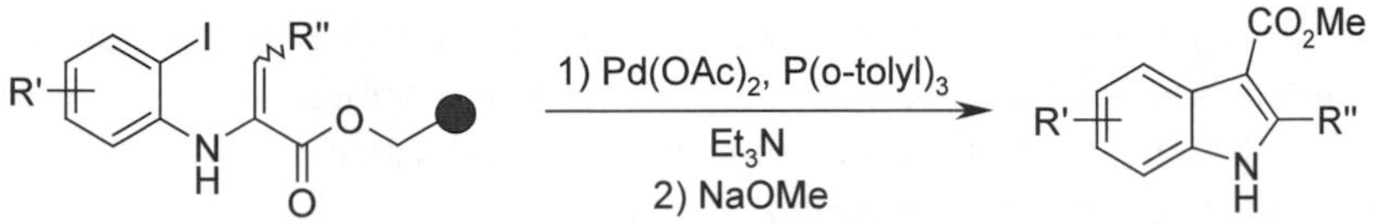

(3.18.)

The pyrrole ring can also be constructed starting from an *N*-vinyl-2-halobenzoic amide. The *N*-(2-iodobenzoyl)-1,4-dihydropyridine derivative shown in **3.19.** underwent palladium catalysed ring closure to give a condensed isoindolone derivative. The use of formic acid as co-solvent led to the reduction of the intermediate palladium complex formed in the insertion step, instead of β-hydride elimination. The transfer of the stereochemical information from the starting material to the product was poor.[25]

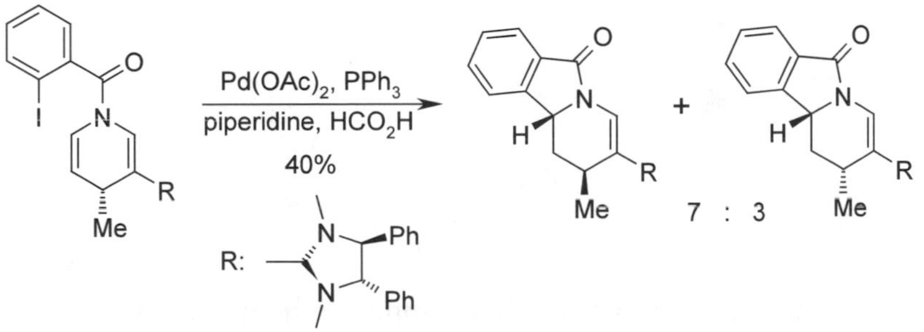

(3.19.)

N-(*Z*-3'-bromoacryloyl)-1-aminocycloalkenes were subjected to Heck conditions to give spiropyrrolones in varying yields, depending on the size of the spiro-fused rings (**3.20.**).[26]

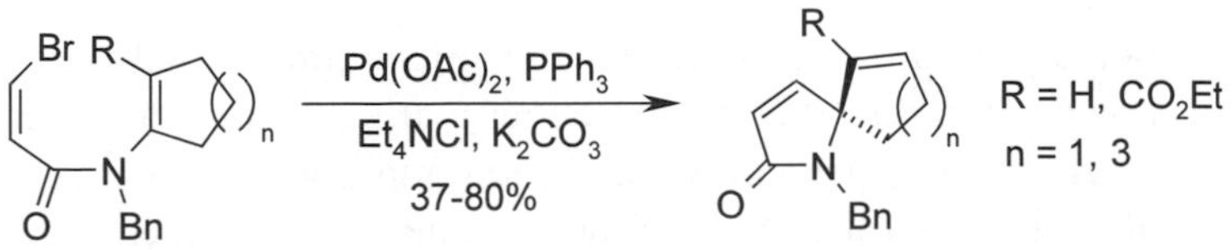

(3.20.)

The Heck reaction was also efficient in the construction of furane and thiophene rings. These processes start form the appropriate *o*-halophenyl ether or sulphide. 2-Butenyl-(2'-iodophenyl)-sulfide, for example, was converted to 3-methyl-benzothiophene under standard conditions in 70% yield (**3.21.**).[27]

In an analogous reaction ethyl *β*-(2'-bromphenoxy)-acrylate underwent ring closure to give the appropriate benzofurane derivative (**3.22.**).[28] The same transformation was the key step in the preparation of benzofuranes on solid support, enabling the generation of benzofurane libraries by parallel or combinatorial synthesis. Cyclization of the resin bound intermediate shown in **3.23.** under "ligand-free" conditions, followed by double bond migration and the detachment of the product from the resin delivered the benzofurane derivatives in good yield and excellent purity.[29]

$$\begin{array}{ccc} & \text{Pd(PPh}_3)_4 & \\ & \xrightarrow{\text{Et}_3\text{N, MeCN}} & \\ & 70\% & \end{array} \qquad (3.21.)$$

$$\begin{array}{ccc} & \text{Pd(OAc)}_2, \text{PPh}_3 & \\ & \xrightarrow{\text{NaHCO}_3, \text{DMF}} & \\ & 70\% & \end{array} \qquad (3.22.)$$

$$\begin{array}{ccc} & 1) \text{ Pd(PPh}_3)_2\text{Cl}_2 & \\ & \text{Et}_3\text{N, Bu}_4\text{NCl} & \\ & \xrightarrow{} & \\ & 2) \text{ TFA, DCM} & \\ & R = H, 83\% & \\ & R = 5,7\text{- di-Cl}, 81\% & \end{array} \qquad (3.23.)$$

Since the benzofurane furan ring is a common structural motif in natural products it's not surprising that the intramolecular Heck reaction has also been employed towards that goal. Such variants that introduce a new centre of chirality into the product are of particular importance.

Members of the retinoid natural product class, containing a quaternary benzylic stereocenter, were prepared by the ring closure of allyl-arylethers. The palladium catalysed process in the presence of sodium formate led to the formation of a partially reduced benzofurane derivative through an insertion-hydride capture reaction sequence (**3.24.**). The coupling, which was run in the presence of a silver-zeolite, showed acceptable enantioselectivity when (*R*)-BINAP was used as ligand.[30]

X = CO$_2$Me, Y = H,   42% (81% *ee*)
X = H, Y = CO$_2$Me,   56% (69% *ee*)

(3.24.)

The carbon skeleton of natural product galanthamine was prepared by the ring closure of an enantiopure aryl-cyclohexenyl ether. The transformation, run in the presence of a palladium-dppp catalyst and silver carbonate, led to the diastereoselective formation of the tricyclic product (**3.25.**). It is worth mentioning, that the chiral information in the ether intermediate was introduced in the palladium catalysed Tsuji-Trost reaction (*N.B.* the synthetic approach depicted in **3.13.** and **3.25.** are much alike).[31]

(3.25.)

Annulation through acetylene insertion

The transition metal catalysed formation of five membered heterocycles through the insertion of a triple bond has also been explored. *o*-Halophenyl-alkynylamines, propargylamines and propargyl-ethers have been subjected to ring closure reactions. These processes, however also require the presence of a second, "anionic" reagent, which converts the palladium complex formed in the insertion step to the product.

The protected *N*-(2'-butynyl)-o-iodoaniline derivative shown in **3.26.** was converted to the corresponding indole derivative. Running the reaction in the presence of organozinc reagents, a series of 3-alkylidenedihydroindoles were isolated in good yield.[32] The analogous reaction of the *N*-propargylaniline derivative in the presence of norbornene resulted in cyclopropane formation and the shift of the double bond to give an indole derivative.[33]

The palladium catalysed reductive insertion of acetylenes is also viable through the use of formate ions as hydride equivalent (*c.f.* **3.24.**). The ring closure of *N*-acetylenic-2-iodoanilines gave the corresponding 3-alkylideneindolines in a selective manner (**3.27.**), demonstrating that the insertion reaction (carbopalladation) is *syn*-selective and the concluding steps of the catalytic cycle preserve this geometrical information.[34]

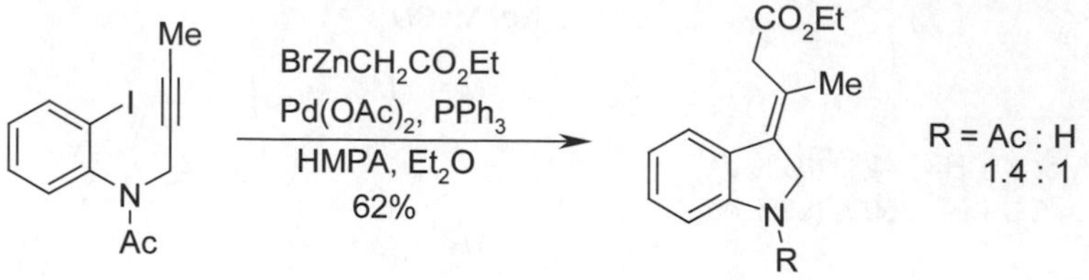

$$(3.26.)$$

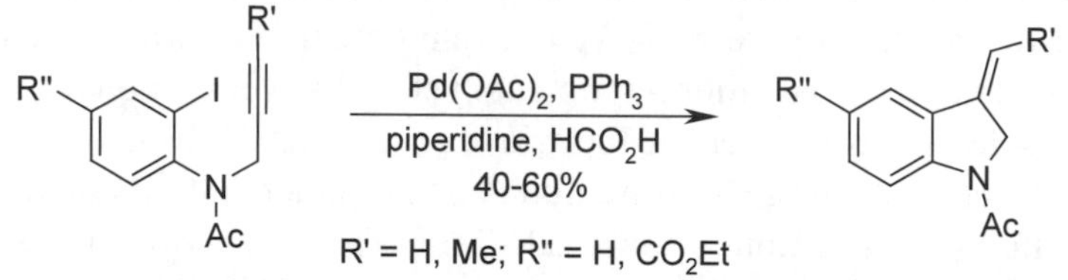

$$(3.27.)$$

Indoles are accessible through the ring closure of *N*-ethynyl-2-haloaniline derivatives too. The capture of the formed indolylpalladium intermediate through coupling with secondary amines in the presence of a suitable base led to the formation of 2-aminoindole derivatives in excellent yield (**3.28.**). The preparation of the starting material was achieved by the *N*-alkynylation of the appropriate aniline derivative using alkynyliodonium salts.[35]

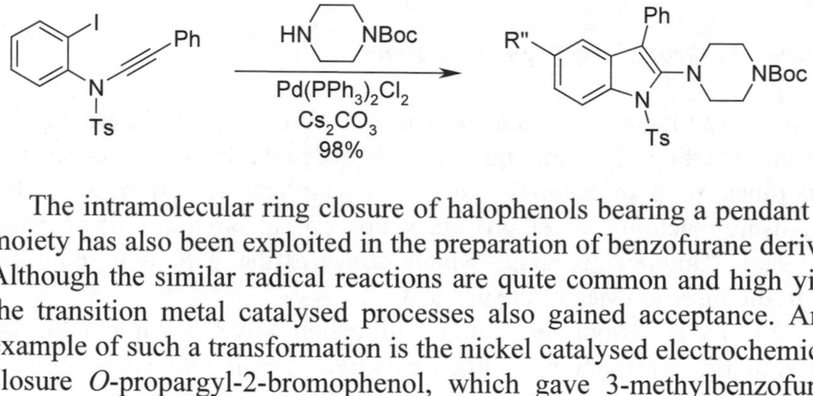

$$(3.28.)$$

The intramolecular ring closure of halophenols bearing a pendant alkyne moiety has also been exploited in the preparation of benzofurane derivatives. Although the similar radical reactions are quite common and high yielding, the transition metal catalysed processes also gained acceptance. An early example of such a transformation is the nickel catalysed electrochemical ring closure *O*-propargyl-2-bromophenol, which gave 3-methylbenzofurane in mediocre yield (**3.29.**).[36]

Most of the ring closure reactions of propargylphenols bear resemblance to the formation of indole derivatives from propargylanilines. Starting from *O*-propargyl-2-iodophenol in the presence of a palladium catalyst and phenylzinc chloride (**3.30.**) led to the formation of the 3-alkylidenedihydrobenzofurane in good yield (*c.f.* **3.26.**).[37] The use of norbornene as capture reagent under similar conditions led to the formation of benzofurane bearing a tetracyclic hydrocarbon substituent.[38]

The addition of allene and a secondary amine to *O*-propargyl-2-iodophenol led to the sequential intramolecular insertion of the acetylene bond, followed by the intermolecular insertion of allene and finally the capture of the formed allylpalladium complex by the secondary amine to give benzofurane analogues bearing an exocyclic diene system (**3.31.**).[39]

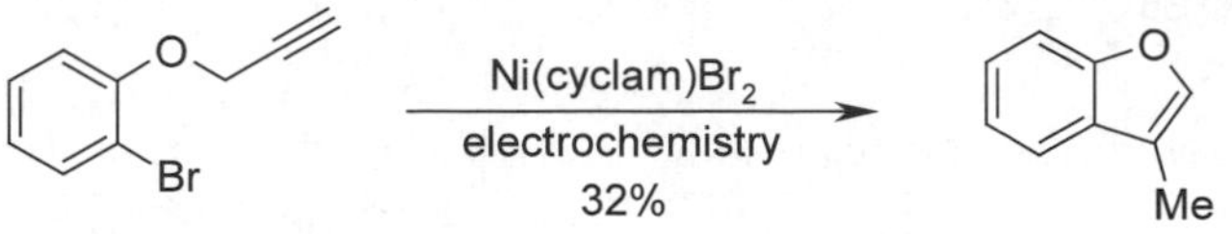

(3.29.)

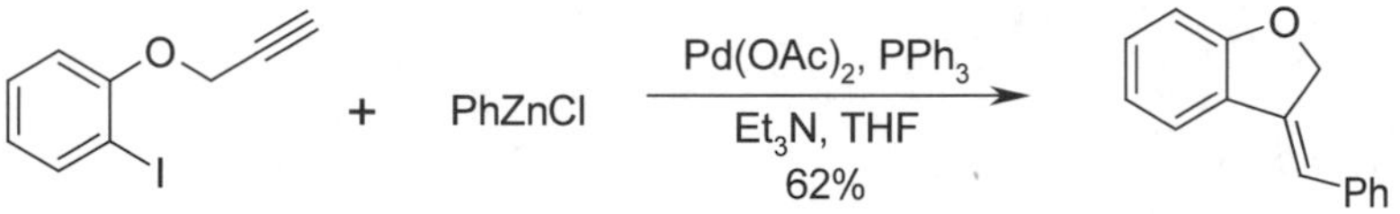

(3.30.)

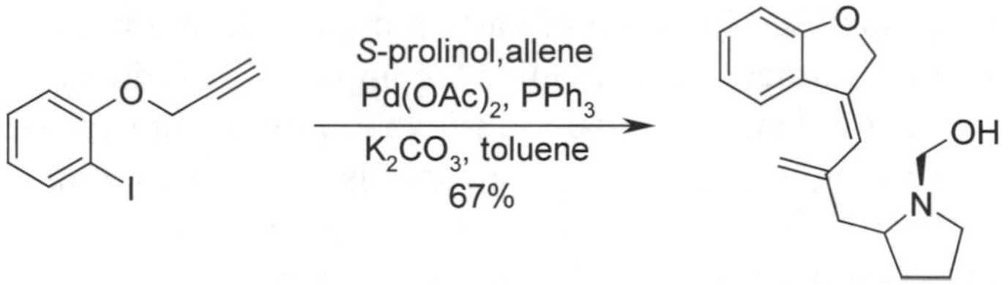

(3.31.)

Examples of the formation of sulphur containing heterocycles by similar transformations are quite rare. The palladium catalysed ring closure of S-propargyl-2-iodophenylmercaptane in the presence of formic acid and an amine-base, analogously to phenols and anilines, led to the formation of the partially reduced benzothiophene skeleton, bearing a 3-methylydene function (**3.32.**).[40]

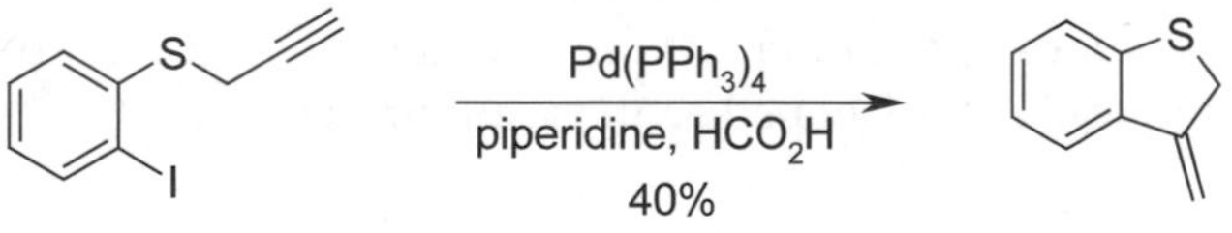

(3.32.)

An analogue of the above transformations is the ring closure depicted in **3.33.**, which involves the formal addition of the phenylpalladium complex formed in the opening step of the catalytic cycle onto the carbon-nitrogen triple bond. In the above process *o*-methylamino-benzonitrile is also formed as by-product in 38% yield, indicating the presence of competing pathways.[41]

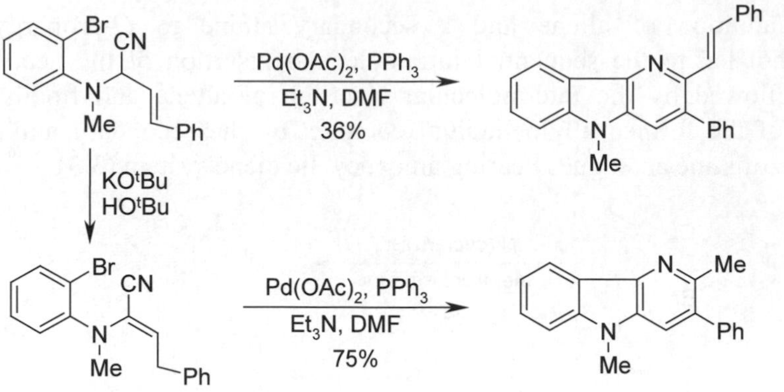

(3.33.)

CO-insertion

2-Bromobenzyl alcohol and its derivatives were converted to phthalides by the palladium catalysed insertion of carbon monoxide and intramolecular quenching of the formed acylpalladium complex. 2-Hydroxymethyl-1-bromonaphthaline, for example, gave the tricyclic product in excellent yield (**3.34.**). An interesting feature of the process is the use of molybdenum hexacarbonyl as carbon monoxide source. The reaction was also extended to isoindolones, phthalimides and dihydro-benzopyranones.[42]

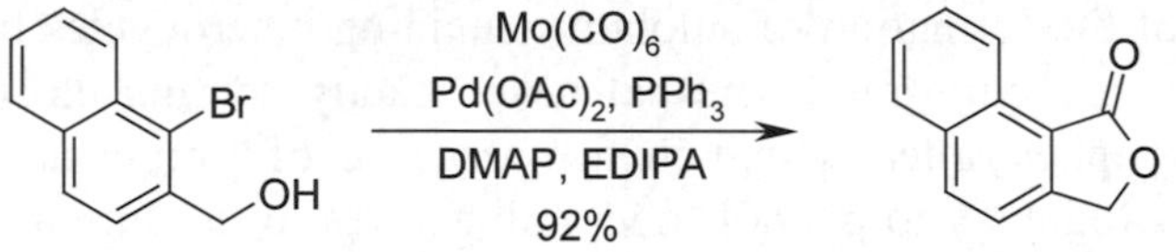

(3.34.)

Furane derivatives were also prepared by the carbonylation of acetylene derivatives. Phenylacetylene was converted to the furanone derivative shown in **3.35.** under reductive conditions, while in the presence of oxygen 2-phenylmaleic anhydride was isolated as the main product.[43]

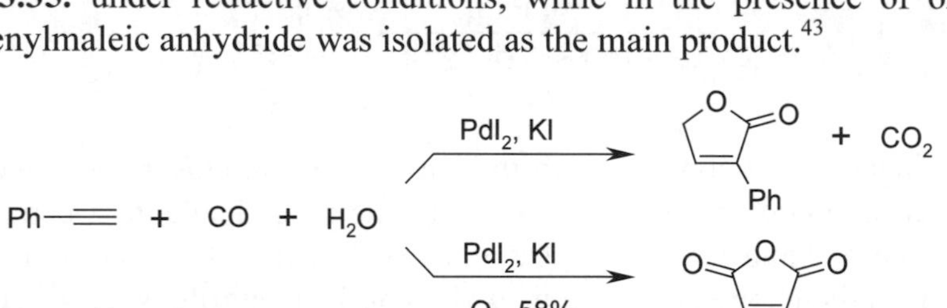

(3.35.)

In a similar double carbonylation process 2-(propargyl)allyl phosphonates were converted into condensed, unsaturated γ-lactones under anhydrous conditions (**3.36.**).[44]

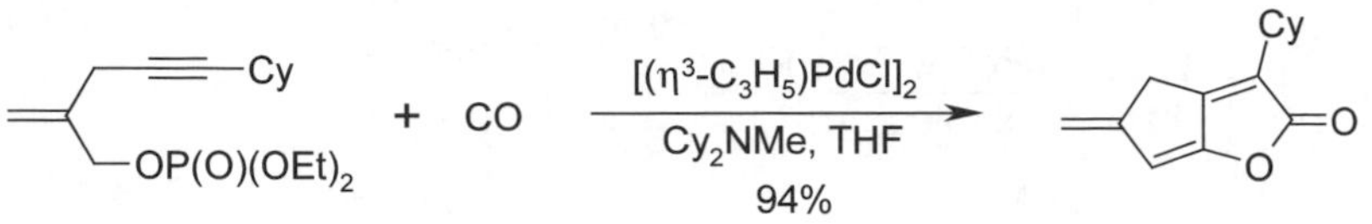

(3.36.)

3.3 CARBON-HETEROATOM BOND FORMATION

A significant part of the examples of transition metal catalyzed formation of five membered heterocycles utilizes a carbon-heteroatom bond forming reaction as the concluding step. The palladium or copper promoted addition of amines or alcohols onto unsaturated bonds (acetylene, olefin, allene or allyl moieties) is a prime example. This chapter summarises all those catalytic transformations, where the five membered ring is formed in the intramolecular connection of a carbon atom and a heteroatom, except for annulation reactions, involving the formation of a carbon-heteroatom bond, which are discussed in Chapter 3.4.

Buchwald-Hartwig coupling

2-Iodophenethylamine derivatives were found to undergo palladium catalysed cyclization to indolines.[45] Starting from the appropriate tetrahydro-isoquinoline derivative the reaction was extended by Buchwald to the preparation of tricyclic natural products (**3.37.**).[46]

Members of the tetracyclic dibenzopyrrocoline alkaloid family can be prepared by the intramolecular ring closure of 1-(*o*-halobenzyl)-tetrahydroisoquinoline derivatives. (**3.38.**).[47] The analogous transformation of dihydroisoquinolines (**3.39.**) proceeds probably through the isomeric enamine form obtained by the tautomeric shift of the double bond.[48] The palladium-carbene catalyst system applied in these reactions was also effective in the preparation of indoline, indolizidine and pyrrolizidine derivatives.[49]

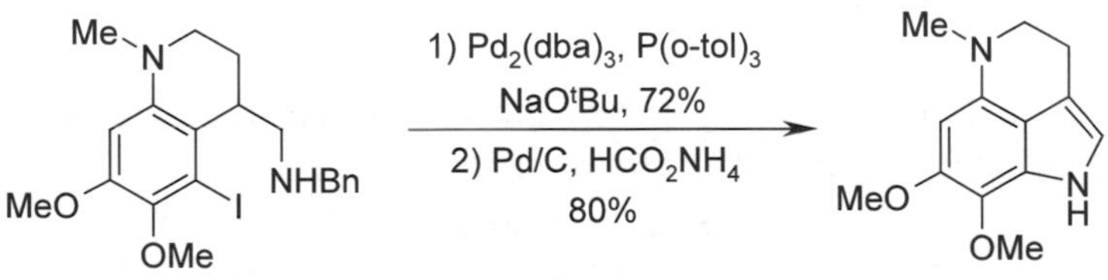

(3.37.)

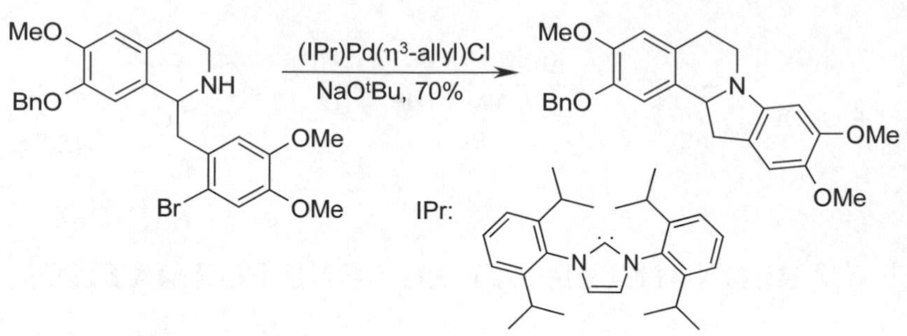

(3.38.)

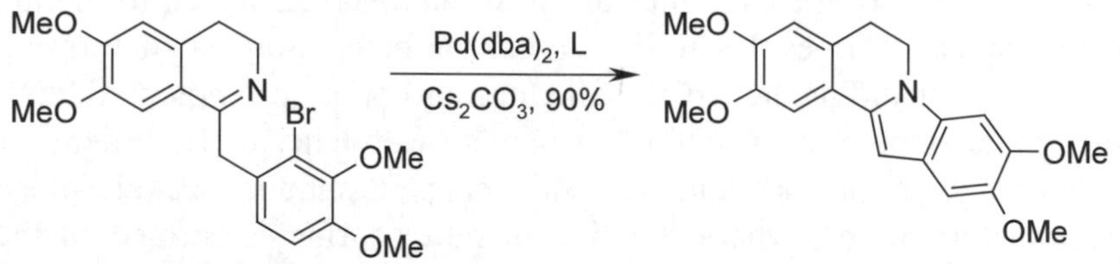

(3.39.)

The formation of two carbon-nitrogen bonds at the same nitrogen atom, using 2,2'-dihalobiphenyls was utilised in the formation of carbazole derivatives. This strategy was also extended to heterocyclic carbazole analogues, such as dithienopyrrole, although with varying success (**3.40.**). Sterically hindered aryl amines, such as diamino-binaphthyl were also transformed efficiently to the corresponding bis(carbazolyl)-biphenyls.[50]

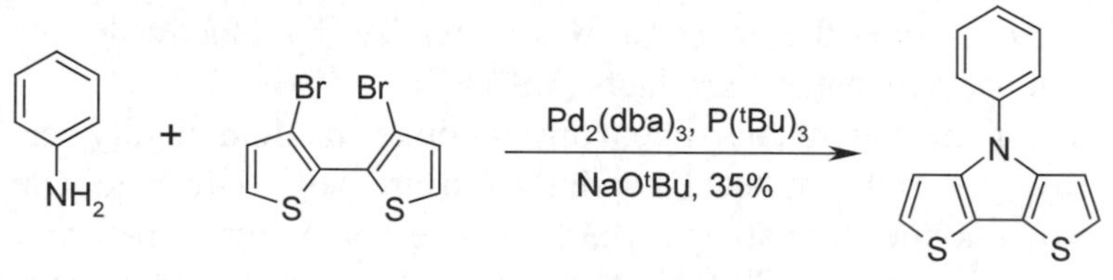

(3.40.)

The palladium catalysed formation of indole derivatives in intramolecular cyclization was also achieved using supported reagents. Polymer bound enamides were reacted with 2-bromo-iodobenzene to get indole derivatives in a Heck coupling-Buchwald-Hartwig coupling sequence (**3.41.**). Removal of the product from the resin was also achieved in the same operation, giving indole-2-carboxylic ester in good yield. The use 2-bromobenzaldehyde or methyl 2-bromobenzoate as starting material led to the formation of isoquinoline or isoquinolone derivatives.[51]

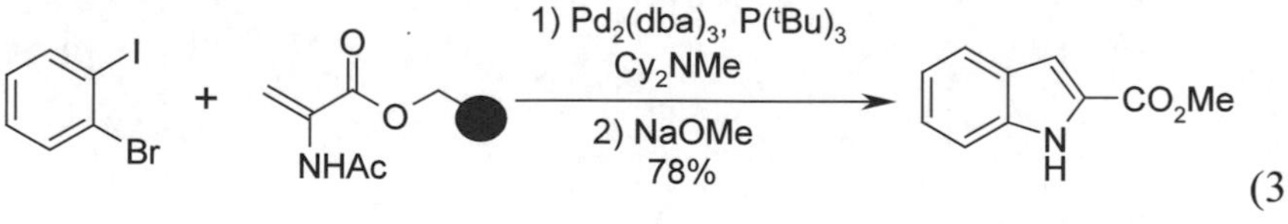

(3.41.)

Copper was also effective in the creation of carbon-nitrogen bonds. *N*-(*o*-halophenyl)-guanidines gave 2-aminobenzimidazoles in the presence of a copper-1,10-phenantroline catalyst system. While the iodo and bromo derivatives in the presence of the copper catalyst gave the cyclized product in 83% and 96% yields respectively, the analogous palladium catalysed transformation showed similar efficiency in both cases, resulting in 88% and 86% yields.[52]

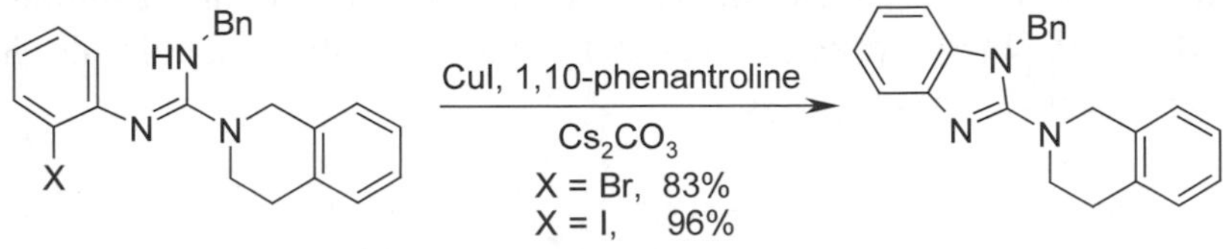

(3.42.)

The formation of oxygen heterocycles through carbon-oxygen bond formation was also reported. Substituted 2-(*o*-halophenyl)-ethanols were converted to dihydrobenzofuranes using palladium and Buchwald's bulky biaryl-type ligands (**3.43.**). The reaction was also efficient in the formation of six and seven membered oxygen heterocycles.[53]

The conversion of 2-(*o*-chlorophenyl)-ethanol to dihydrobenzofurane was also promoted by copper(I) chloride in toluene in the presence of sodium hydride (**3.44.**). The choice of solvent was crucial to suppress undesired side reactions, such as dehydration or the loss of a formaldehyde fragment.[54]

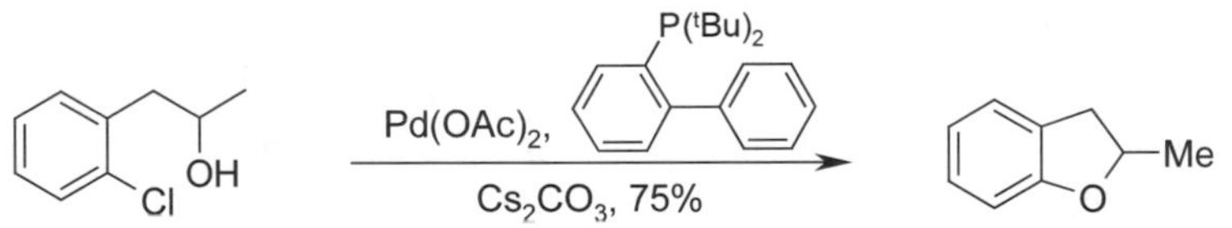

(3.43.)

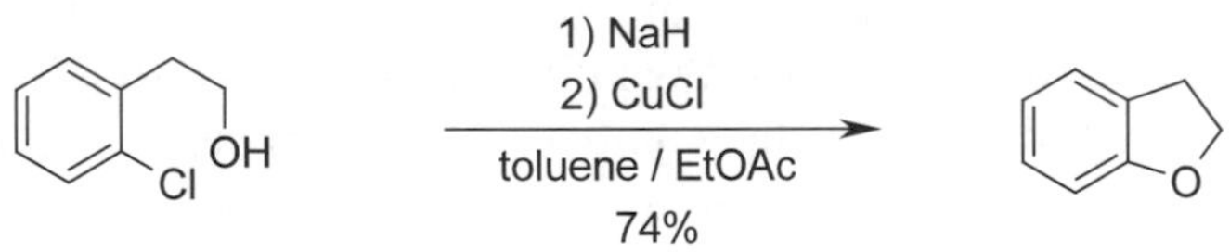

(3.44.)

Addition onto acetylenes

The transition metal mediated addition of N-H or O-H bonds onto triple bonds is a common reaction, which is extensively used in the preparation of five membered rings, indoles and benzofuranes being the most frequent targets. The transformation is not restricted to catalytic processes and the use of a strong base might lead to the same result.[55]

A characteristic example of these reactions is the conversion of propargylgylicine to the dehydroproline derivative shown in **3.45**. Unfortunately under efficient coupling conditions the enantiopurity of the product is only mediocre due to partial racemization, while the conservation of the optical purity is accompanied by mediocre yields.[56]

The ring closure of aminoalkynes bearing a leaving group in the appropriate position might lead to the formation of pyrroles in an addition-elimination sequence. 2-Phenylethynyl-1-amino-*sec*-butanol, for example, gave 4-ethyl-2-phenylpyrrole on treatment with palladium dichloride in acetonitrile in excellent yield (**3.46.**).[57]

(3.45.)

(3.46.)

The most common application of the intramolecular addition of amines onto carbon-carbon triple bonds is the preparation of indoles from *o*-ethynyl-anilines. The reaction might be carried out in a myriad of ways, of which we only show some typical examples.

A series of 2-ethynylaniline derivatives were prepared in a consecutive Sonogashira coupling sequence starting from 2-iodoaniline and trimethylsilylacetylene. Their ring closure in the presence of palladium dichloride led to the formation of the corresponding indole derivatives in good yield (**3.47.**).[58] Variations of this approach include the formation of the ethynylaniline derivative from *o*-haloaniline and a tin-acetylide (Stille coupling)[59] or from a thalliumorganic compound and a copper acetylide.[60]

In certain cases the presence of a free amino group leads only to moderate yields. The purine analogue, prepared in 62% in Sonogashira coupling from 2-ethynylaniline underwent copper catalysed ring closure to give the desired product in only 6% yield, while its *N*-mesityl derivative not only gave higher yields in the Sonogashira coupling, but also underwent spontaneous ring closure under the coupling conditions to give the indole derivative in 78% yield (**3.48.**).[61] The copper catalysed formation of indoles from *o*-ethynylanilines was also exploited in the microwave assisted solid-phase synthesis of indole libraries.[62]

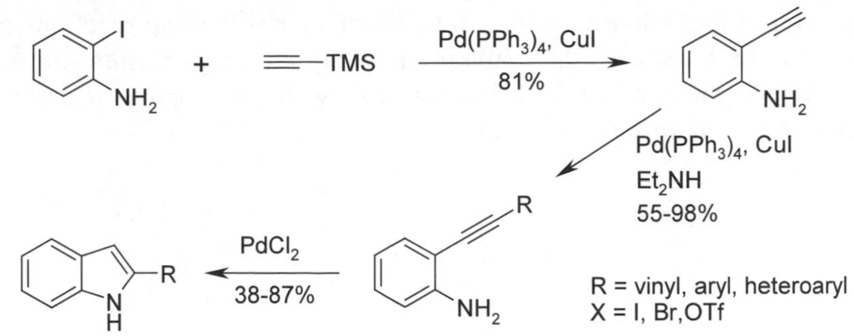

(3.47.)

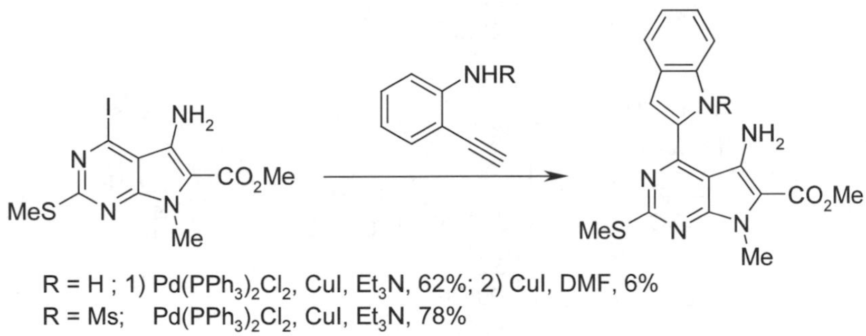

R = H ; 1) Pd(PPh₃)₂Cl₂, CuI, Et₃N, 62%; 2) CuI, DMF, 6%
R = Ms;   Pd(PPh₃)₂Cl₂, CuI, Et₃N, 78%

(3.48.)

The preparation of indoles was accomplished from arylacetylenes and *N*-protected-2-iodoanilne under solvent-free conditions in the presence of KF doped alumina and a palladium-copper catalyst system, using microwave irradiation to promote the transformation. In the high yielding process the ratio of the Sonogashira coupling product and the ring closed indole derivative depended on the *N*-substituent, mesyl and acetyl favouring ring closure (**3.49.**). Using similar conditions the methodology was successfully extended to the preparation of 2-arylbenzofurane derivatives too.[63]

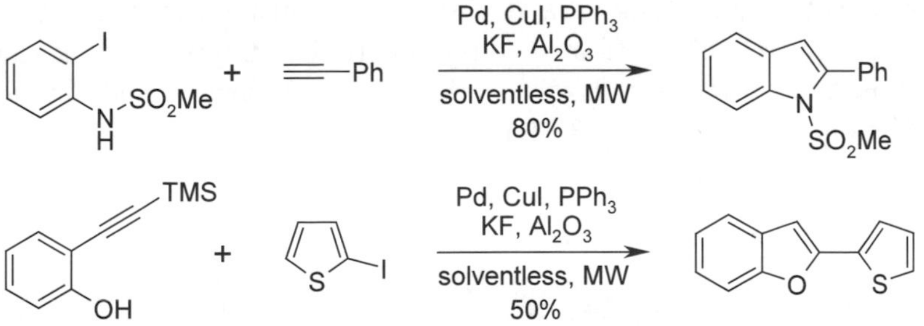

(3.49.)

The intramolecular addition of the N-H bond onto the acetylene moiety in ethynylanilines is presumed to lead to the formation of an indolylpalladium complex (*c.f.* Chapter 1.2.). Trapping of this intermediate in subsequent transformations leads to the introduction of a substituent into the 3-position of the indole ring. A collection of such follow-up reactions is presented in **3.50.**, where the palladium catalysed ring closure of *N*-trifluoroacetyl-2-phenylethynyl-aniline was followed by coupling reactions leading to a series of various indole derivatives.[64]

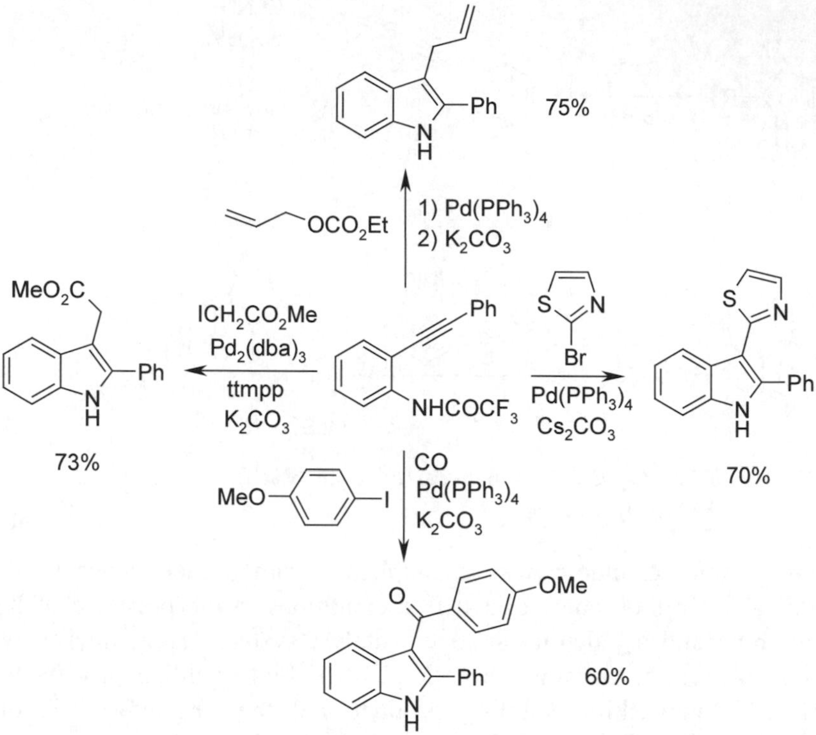

(3.50.)

In spite of the fact, that it is accomplished by the attack of a nitrogen atom on a triple bond, the copper catalysed ring closure of (*o*-ethynylphenyl)-triazenes follows a different mechanistic pathway. In the presence of copper(I) chloride in dichloroethane the starting triazenes are converted into *N*-amino-isoindazoles, while their thermal transformation gives predominantly cinnolines (**3.51.**).[65]

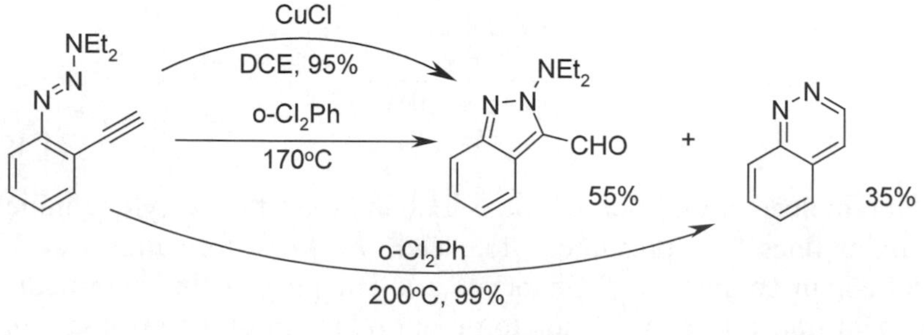

(3.51.)

The intramolecular addition of hydroxyl groups to triple bonds might be utilised in the formation of furane derivatives. Enynols, having the appropriate double bond geometry, underwent ring closure and subsequent double bond isomerisation in the presence of both palladium and ruthenium catalysts to give substituted furans (**3.52.**).[66]

A similar procedure was reported by Quing, where intramolecular addition of an alcohol onto an enyne moiety, followed by the isomerisation of the exocyclic double led to furane formation (**3.53.**).[67]

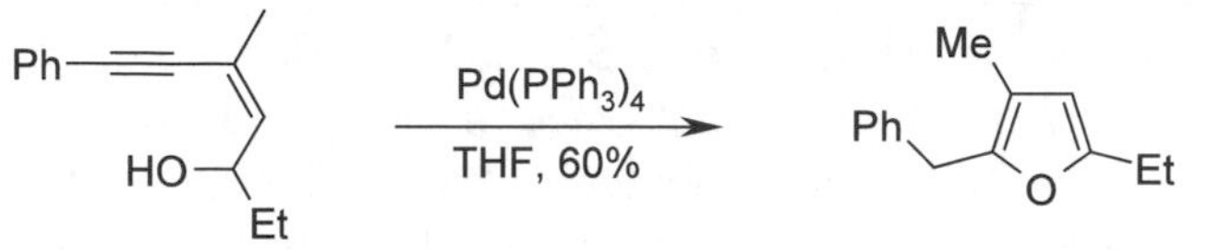

(3.52.)

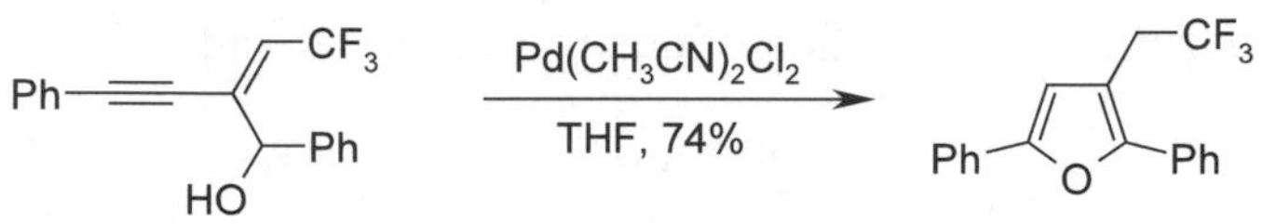

(3.53.)

Unlike certain *o*-ethynylanilines, *o*-ethynylphenols usually undergo spontaneous ring closure to the corresponding benzofurane derivative. The reaction of 2-iodophenol and different arylacetylenes in the presence of a polymer supported palladium catalyst and caesium hydroxide, for example, gave 2-arylbenzofuranes exclusively (**3.54.**).[68] The reaction has also been successfully employed in the preparation of substituted benzofuranes,[69] and benzofurane derivatives on solid support.[70]

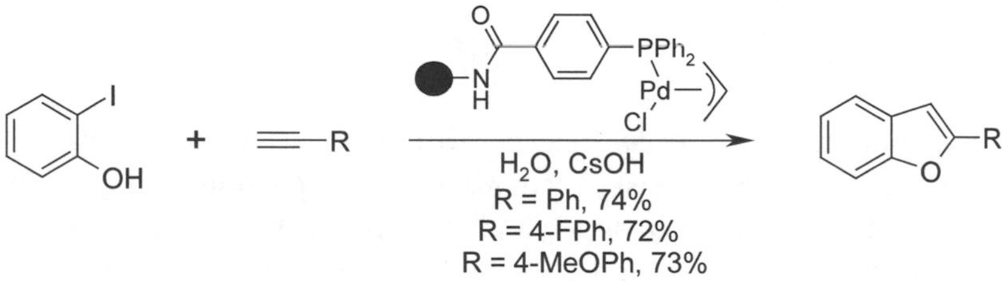

(3.54.)

The same reaction, the conversion of 2-iodophenol to 2-arylbenzofuranes was also accomplished in the presence of a copper-phenantroline catalyst and caesium carbonate. Different arylacetylenes and iodophenols were coupled to give the corresponding benzofurane derivatives in good to excellent yield (**3.55.**)[71]

Cu(phen)(PPh$_3$)$_2$NO$_3$
Cs$_2$CO$_3$, toluene

R = Ph, 92%
R = 4-CNPh, 77%
R = 4-MeOPh, 62%

(3.55.)

Scope of this synthetic strategy is not limited to benzofurans. The reaction of 2-iodo-3-hydroxypyridine and 1,1-diethoxy-2-propyne under Sonogashira coupling conditions (palladium-copper catalyst system) leads to the formation of the substituted furo[3,2-*b*]pyridine shown in **3.56**.[72]

The palladium catalysed coupling of trimethylsilylacetylene and the iodopyridone derivative shown in **3.57.** led to the formation of the corresponding ethynylpyridine. The absence of the spontaneous ring closure to furo[2,3-*b*]pyridine might be attributed to the preference of the pyridone form over the hydroxypyridine form. Addition of a copper catalyst and base, however, led to the closure of the furan ring (**3.57.**).[73] The same approach was utilised in the functionalization of uracil based nucleosides.[74]

Pd(PPh$_3$)Cl$_2$, CuI

piperidine
82%

(3.56.)

Pd(OAc)$_2$, PPh$_3$

CuI, BuNH$_2$
81%

CuI, Et$_3$N

K$_2$CO$_3$, EtOH
84%

(3.57.)

The preparation of (poly)hydroxybenzofuranes, a common motif in natural products, in the Sonogashira coupling of halophenols and acetylene derivatives usually proceeds in low yield, unless the hydroxyl groups are protected. 4-Bromoresorcine[75] and 2-bromohydroquinone[76] were converted into hydroxybenzofuranes in a one-pot protection-coupling-deprotection-ring closure sequence (**3.58.**). Of the protecting groups tested acetyl was the optimal, since it was easy to introduce and remove, and increased the reactivity of the aromatic ring in the Sonogashira coupling.

The same strategy was successfully employed in the preparation of the natural product cicerfurane. The procedure started from 4-bromoresorcine and the sesamol-derived acetylene derivative in **3.59**.[75] One-pot acetyletion, cross-coupling and deprotection followed by spontaneous ring closure led to the isolation of the natural product. The mediocre yield was attributed to the sensitivity of the intermediate electron rich diarylacetylene. Attempts at the preparation of cicerfurane in a stepwise manner led to inferior results.

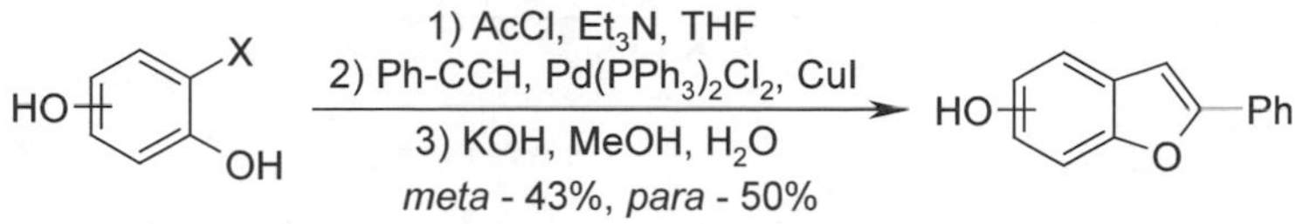

(3.58.)

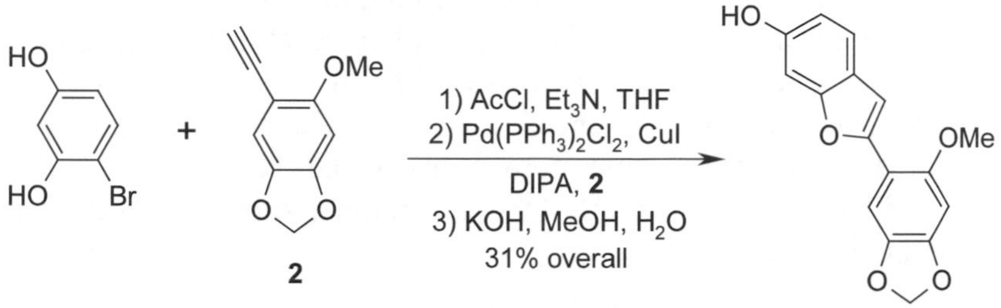

(3.59.)

In analogy with **3.50.** the intermediate benzofuranyl-palladium complex formed in the ring closure might be further elaborated by the appropriate choice of the reaction conditions. The ring closure of 2-phenylethynyl-phenol, if carried out in the presence of 4-iodoanisole might lead to the formation of a 2,3-disubstituted furane derivative as by-product. By the correct choice of the coupling conditions, using a palladium-bipy catalyst, this pathway became dominant, leading to the formation of 3-anisyl-2-phenyl-benzofurane in good yield (**3.60.**).[77]

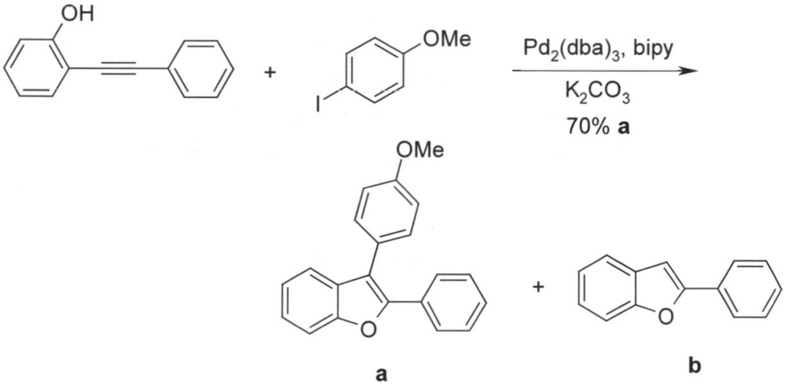

(3.60.)

The same reaction, run in the presence of carbon monoxide, led to the incorporation of an acyl moiety into the 3-position. Ring closure of the ethynylphenol derivative shown in **3.61.** in the presence of iodobenzene and carbon monoxide gave the corresponding 3-anisoyl-benzofurane derivative in excellent yield.[78]

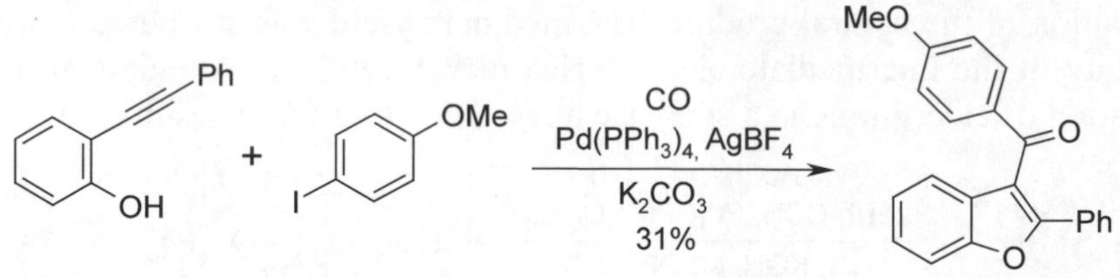

(3.61.)

Unlike in the case of the preparation of indoles and benzofuranes, the synthesis of benzothiophenes from *o*-ethynyl-thiophenols is not known. A close analogy was reported by Larock, where phenylacetylene was coupled with 2-iodothioanisole. Ring closure of the formed *o*-ethynyl-sulfide was initiated by the addition of different electrophiles. The reaction led to the formation of the benzothiophene core bearing the electrophile in the 3-position (**3.62.**). Typical examples include iodine, bromine, NBS, and phenylselenyl chloride.[79]

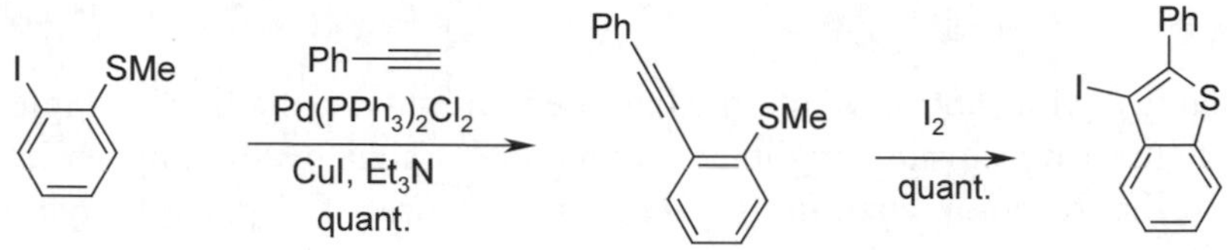

(3.62.)

Reaction with olefins

The palladium catalysed substitution reaction of allylic systems has also been utilised in the formation of five membered rings. Intramolecular nucleophilic attack of the amide nitrogen atom on the allylpalladium complex formed in the oxidative addition of the allyl acetate moiety on the catalyst led to the formation of the five membered ring (**3.63.**). In the presence of a copper(II) salt the intermediate pyrroline derivative oxidized to pyrrole.[80]

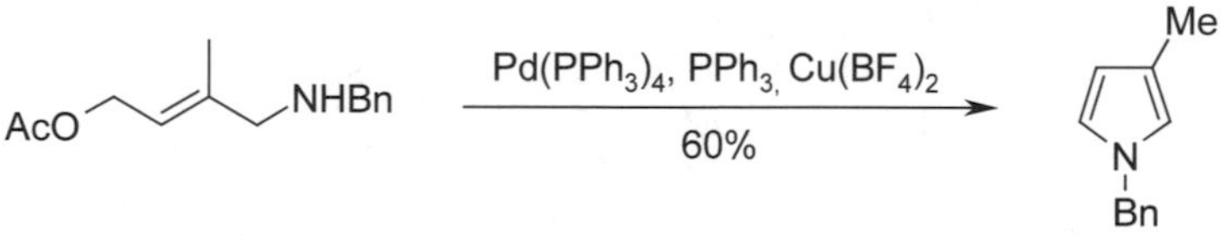

(3.63.)

Although the palladium catalysed conversion of allylsilanes bearing a pendant hydroxyl group to furans (**3.64.**), reported by Szabó, bears remarkable similarity to the reaction depicted in **3.63.**, it has also some distinctive features. The formation of the allylpalladium complex is achieved in a transmetalation step and therefore requires the presence of a palladium(II) catalyst. The re-oxidation of the palladium(0) complex released in the final step of the catalytic cycle was accomplished by the addition of copper(II) compounds.[81]

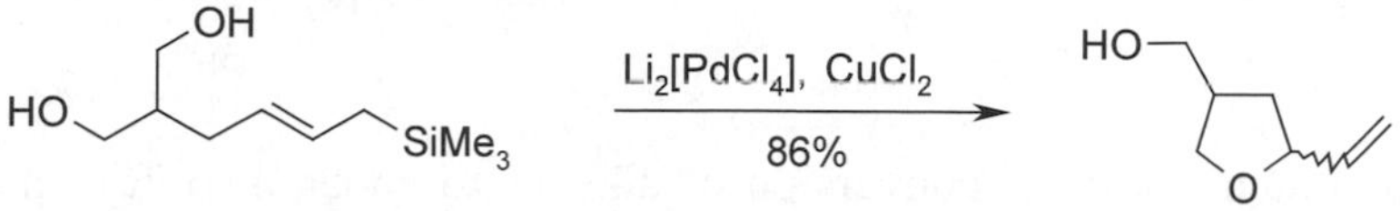

(3.64.)

Not only acetylene derivatives do undergo palladium catalysed intarmolecular carbon-nitrogen bond formation with amines. The similar reaction of olefins in a Wacker-type process also leads to ring closure. ω-Aminopentenes bearing a suitable leaving group in the 4-position were converted to pyrroles in a cyclization-isomerisation-elimination sequence (3.65.).[82]

In an analogous process *o*-allylanilines were converted into indole derivatives by Hegedus (3.66.). Since the process is initiated by a palladium(II) catalyst and produces palladium(0) as product, recycling of the catalyst had to be ensured by the addition of stoichiometric amounts of an oxidant (e.g. benzoquinone).[83]

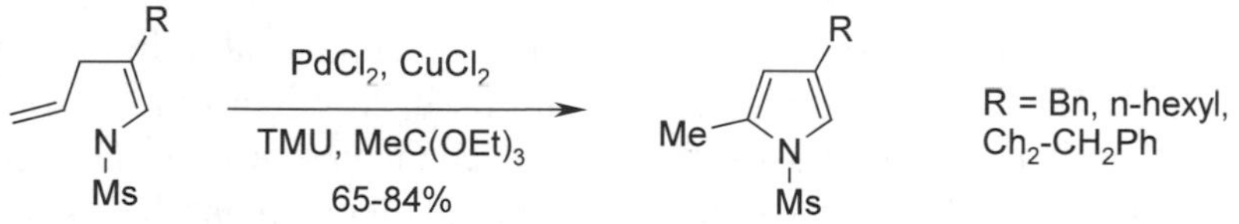

(3.65.)

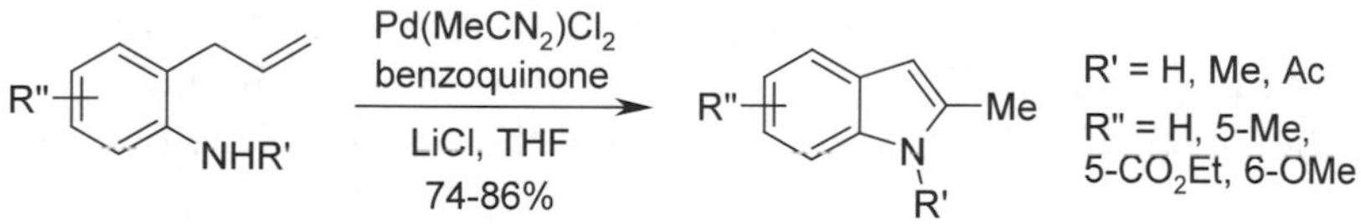

(3.66.)

The combination of the formation of palladium(II) complexes in the oxidative addition of unsaturated halides onto palladium(0), and the palladium(II) mediated ring closure of olefins allows for the elimination of the oxidant. The ring closure of the methallyl-aniline derivative shown in 3.67. in the presence of β-bromostyrene and a catalytic amount of palladium led to the ring closed – coupled product.[84] The ring closure might also be accompanied by other follow-up reactions. 2-Methallyl-anilines were converted for example to indolylacetic acid derivatives in the presence of carbon monoxide.[85]

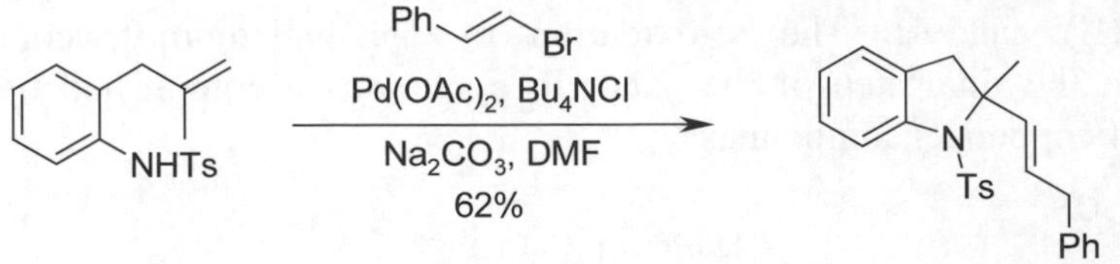

(3.67.)

The palladium catalysed conversion of alkenes to enols, also known as the Wacker reaction, has also been used in the formation of oxygen heterocycles. In the example shown in **3.68.** the subsequent formation of two carbon-oxygen bonds leads to the desired dioxabicyclo[3.2.1]octane derivative. The first Wacker reaction gives selectively a six membered ring formation (other possible routes would lead to even larger rings), while in the second Wacker reaction the selective formation of the five membered ring is observed.[86]

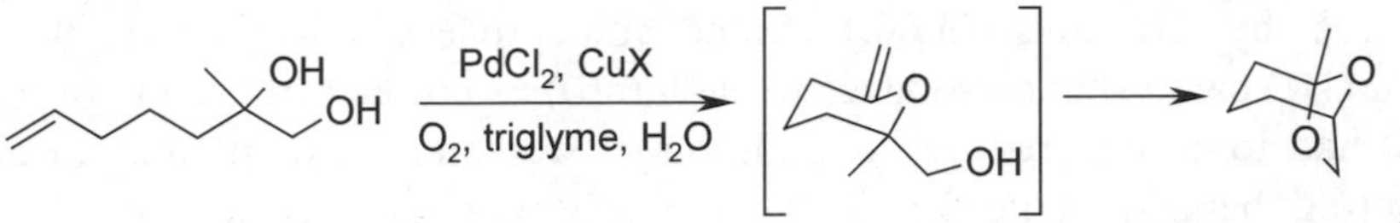

(3.68.)

Alike olefins, allenes also undergo palladium mediated addition in the presence of N-H or O-H bonds. Although these reactions show some similarity to Wacker-type processes, from the mechanistic point of view they are quite different. Allenes, such as the α-aminoallene in **3.69.**, usually undergo addition with palladium complexes (*e.g.* carbopalladation in **3.69.** and **3.70.**, or hydropalladation in **3.71.**), which leads to the formation of a functionalized allylpalladium complex. Subsequent intramolecular nucleophilic attack by the amino group leads to the closure of the pyrroline ring.[87]

The reaction of octylallene and *N*-tosyl-2-iodoaniline (**3.70.**) proceeds analogously, with the only difference that the amino group is born by the aryl iodide and not the allene reagent.[88]

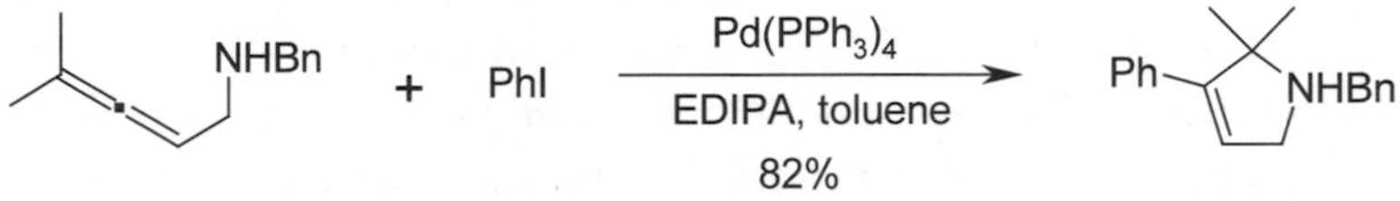

(3.69.)

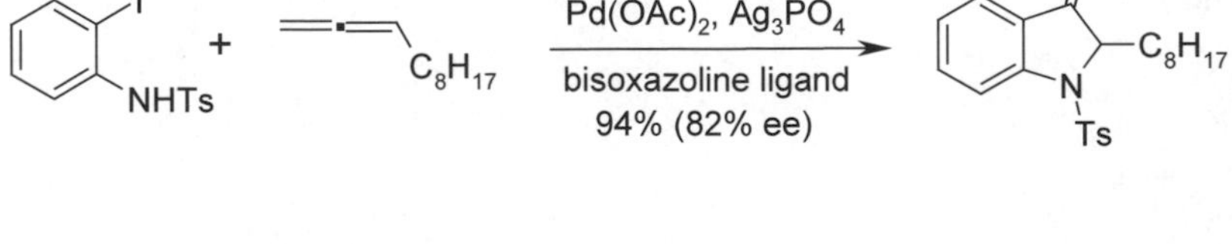

(3.70.)

The palladium catalysed ring closure of 1-phenyl-6-aminohexyne to 2-styrylpyrrolidine proceeds through the formal activation of the propargyl-position. In the presence of benzoic acid palladium(0) is presumably converted into a hydrogenpalladium complex, which isomerises the alkyne to allene, and converts the allene to an allylpalladium complex. Intramolecular attack of the amino function on the allylpalladium complex, as seen before, concludes the ring closure and gives the styrylpyrrolidine (**3.71.**). The process was also effective in the preparation of vinylpiperidines, starting from 7-aminoheptyne derivatives.[89]

(3.71.)

Allenyl ketones and aryl halides undergo coupling and cyclization in the presence of a palladium-triphenylphosphine catalyst and silver carbonate. The reaction leads to the formation of furane derivatives, the aryl group being introduced into the 3-position (**3.72.**).[90]

The formation of 2 furane rings was achieved in one transformation by Ma and co-workers. Allenoic acids and allenyl ketons were reacted in the presence of a palladium catalyst to give the unsymmetrical bifuryl product, arising from the cyclization of both allene derivatives mediated by the same palladium centre followed by their coupling (**3.73.**).[91]

(3.72.)

(3.73.)

Furan and pyrrole derivatives might also be prepared by the copper catalysed isomerisation of alkynyl ketones and their imine derivatives. 1-Acetyl-octyne was for example converted into 1-methyl-5-pentylfurane in the presence of a catalytic amount of copper(I) iodide in excellent yield (**3.74.**).[92] The similar ring closure of ketones bearing a sulphide function in the propargyl position led to the unexpected migration of the thio group into the 3-position of the furane ring.[93]

$$H_{11}C_5 \overset{Me}{\underset{O}{\equiv}} \xrightarrow[\substack{Et_3N,\ DMF \\ 94\%}]{CuI} H_{11}C_5 \overset{O}{\underset{}{\diagdown}} Me$$

(3.74.)

Indole derivatives might also be prepared by the palladium catalysed reductive heteroannulation of *o*-nitrostyrene derivatives. The bicyclic olefin, shown in **3.75.** was converted to the indole derivative in good yield.[94] The reaction, which was run under forcing conditions utilises carbon monoxide as the reducing agent.

The double bond, onto which the ring closure is made, might also be the part of an aromatic ring (although the ring closure in this case proceeds probably by a distinctly different mechanism). 2-Aryl-nitrobenzene derivatives were heated under a slight pressure of carbon monoxide in the presence of a palladium-phenantroline catalyst system to obtain carbazoles and analogous ring systems. 2-Thienyl-nitrobenzene, for example, gave thieno[2,3-*b*]indole in good yield (**3.76.**).The efficiency of the reaction depends both on the source of palladium and the type of ligand used, anilines and urea derivatives being typical side products.[95]

$$\text{(structure, MeO}_2C, CO_2Me) + CO \xrightarrow[65\%]{Pd(OAc)_2,\ dppp} \text{(indole product)}$$

(3.75.)

$$\text{(structure with NO}_2) + CO \xrightarrow[81\%]{Pd(OAc)_2,\ 1,10\text{-phenantroline}} \text{(thieno indole product)}$$

(3.76.)

3.4   OTHER PROCESSES

Besides ring closure reactions proceeding through C-H activation, a major part of this chapter is devoted to the formation of five membered heterocycles in annulation reactions.

The ability of palladium(II) acetate to mediate the oxidative ring closure of indole derivatives containing a pendant aromatic ring has been exploited by Itahara already in the late 70s.[96] An extension of the original work to carbonyl-diindole (**3.77.**) has been reported by Black.[97]

The use of a stoichiometric amount of palladium acetate, a fact that biases the original oxidative ring closure reactions, can be overcome by the use of an oxidant in the process, which re-oxidizes palladium(0) that is formed in the final step of the ring closure. Such a transformation is presented in **3.78.**, where an anilino-benzoquinone was ringd closed to give an indoloquinone in the presence of a catalytic amount of palladium acetate and a stoichiometric amount of copper(II) acetate.[98]

(3.77.)

(3.78.)

N-aryl-*o*-haloanilines can be converted into indole derivatives in a palladium catalysed oxidative addition, C-H activation sequence. The transformation has been utilized extensively in the preparation of polycyclic compounds. In a recent example, leading to the formation of the carbazole ring system, Larock and co-workers demonstrated that the formation of the *o*-halo-diphenylamine derivative and subsequent establishment of the carbon-carbon bond might be carried out in one pot. The carbon-nitrogen bond was formed by the addition of the *o*-haloaniline onto a substituted benzyne generated *in situ*, while the addition of a palladium-tricyclohexylphosphine catalyst system initiated the ring closure (**3.79.**). Starting from *o*-halophenols the process was equally efficient in the formation of dibenzofurans.[99]

Electron deficient systems, such as quinolines, are also susceptible to intramolecular palladium catalysed ring closure. 4-(2'-Chloroanilino)-quinoline, for example gave indolo[3,2-*c*]quinoline in excellent yield (**3.80.**). The dipyridopyrrole skeleton was also prepared in this manner starting from 2,3-dichloropyridine and 4-aminopyridine.[100]

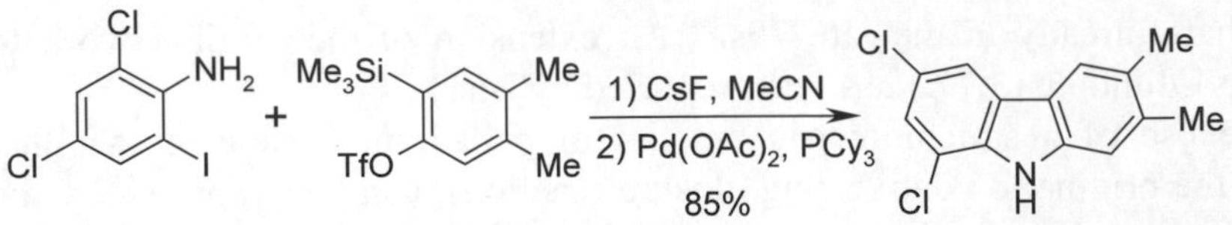

(3.79.)

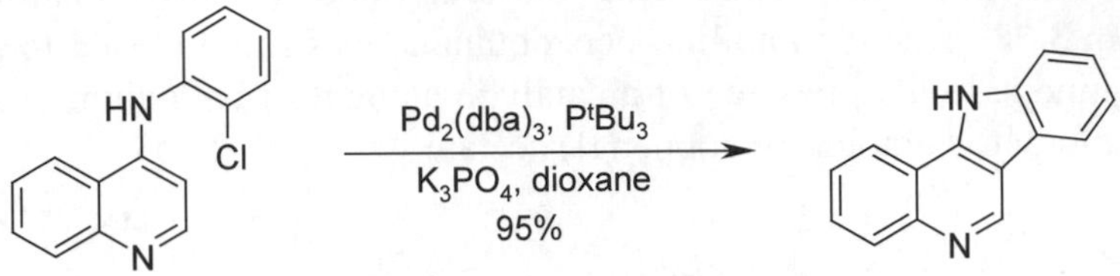

(3.80.)

The same transformation has also been extended to pyridone-type systems, where the ring closure might also be described as a Heck-type coupling. Padwa reported the conversion of condensed 3-(*o*-bromoanilino)-2-pyridone derivatives to *β*-carbolinone (**3.81.**).[101]

The analogous reaction of 5-iodopyridazin-3(2H)-one led to the formation of pyridazino[4,5-*b*]indoles (**3.82.**) in a two step nucleophilic substitution, palladium catalysed intramolecular carbon-carbon bond formation sequence. The same reaction has also been carried out in one-pot.[102] An elegant application of this methodology has been reported by ·Harayama in the preparation of the pentacyclic natural product Luotonin A.[103]

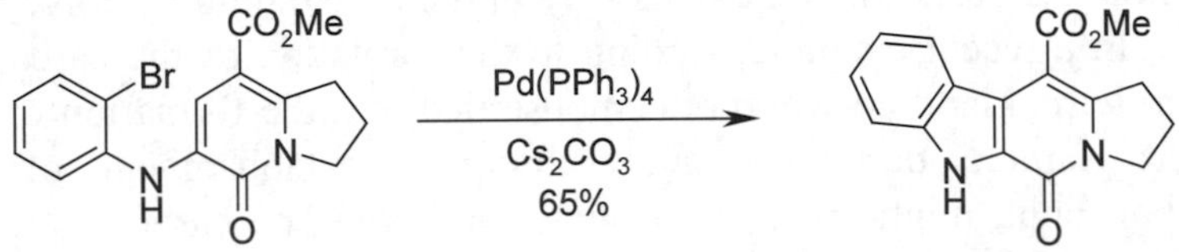

(3.81.)

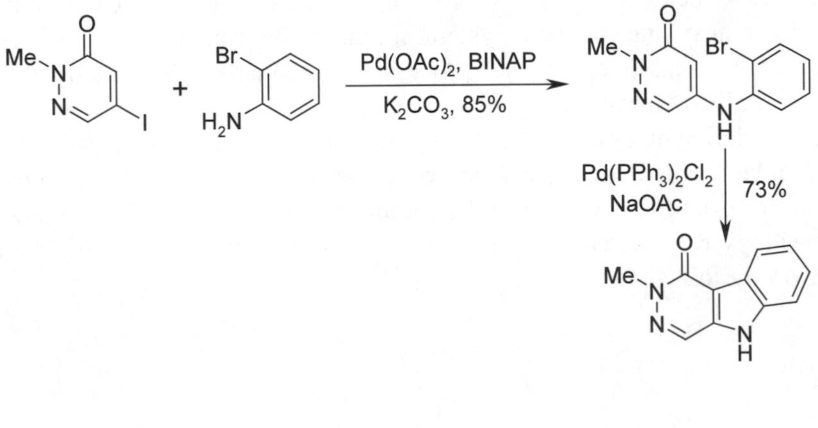

(3.82.)

The palladium catalysed annulation of β-iodoamines and acetylene derivatives has been widely used in the construction of pyrroles and indole derivatives. The strategy, first reported by Larock, is efficient in converting both cyclic and open chain 2-iodo-allylamine derivatives to 3-alkylidene-2,3-dihydropyrrole derivatives (**3.83.**). In these reactions the formation of the carbon-carbon bond precedes the formation of the carbon-nitrogen bond.[104]

(3.83.)

The most frequent application of Larock's method is undoubtedly the conversion of o-haloanilines to indoles. The scope of this transformation covers a wide range of disubstituted acetylenes and anilines including both N-substituted and unsubstituted derivatives. A schematic representation of the process with some selected examples is given in **3.84.**[105] Extension of the transformation to solid phase linked aniline derivatives has been utilized in the construction of 2,3-disubstituted indole libraries.[106] The reaction works also well with o-halophenols giving 2,3-disubstituted benzofuran derivatives as product.[107]

R' = H, Me, Ts
R", R''' = nPr, Cy, TMS, Ph, CH_2OH, C(Me)=CH_2, $(CH_2)_2OH$, CMe_2OH

(3.84.)

The selectivity of the process is demonstrated in the reaction of 2-iodoaniline and the pentenyne derivative shown in **3.85**. The insertion takes place preferentially on the triple bond (*c.f.* **3.87.**), giving rise to the isoporpenylindole derivative in good yield.[105]

(3.85.)

The transformation is not limited to anilines. Other *ortho*-disubstituted aromatic compounds work also well as substrates. The N-Boc protected 2-iodo-3-aminothiophene derivative was transformed into the corresponding pyrrolothiophene in good yield (**3.86.**). The use of a silyl group was also tolerated by the reaction conditions allowing for the introduction of a TBS-group into the product.[108] The analogous 3-iodo-2-aminopyridines gave pyrrolopyridines in varying yield.[109]

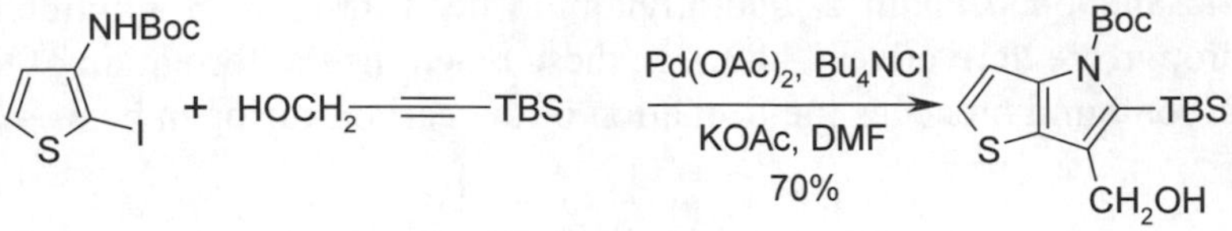

(3.86.)

Larock has also extended the annulation chemistry to olefins. 1,3-Dienes, both acyclic and cyclic, were reacted with iodophenol[110] and analogous compounds to give condensed benzofurans. The 7-acetoxy-8-iodocoumarin derivative shown in **3.87.**, for example, was converted to the corresponding tetracyclic furocoumarin derivative in good yield.[111]

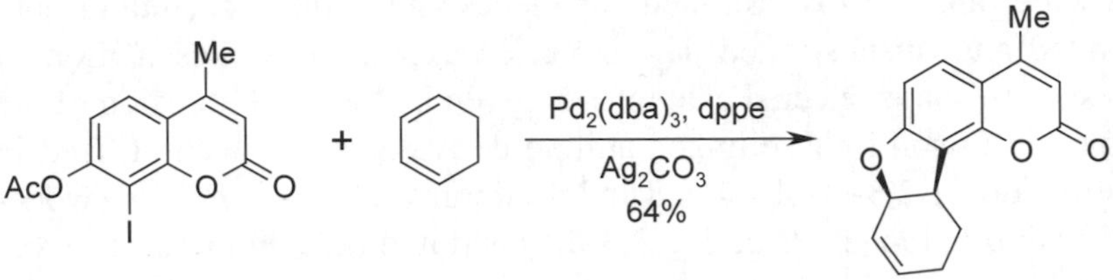

(3.87.)

Although the following reactions (**3.88.-3.93.**) show certain superficial similarity in involving the formation of five membered heterocycles starting from unsaturated carbon-carbon and/or carbon heteroatom bonds, they all have a complex, and sometimes strikingly different mechanism.

Pyrroles and furans were prepared by the intramolecular carbon-carbon bond formation between pendant acetylene and nitrile or carbonyl functions. The process, running in acetic acid, starts by the *trans*-acetoxypalladation of the acetylene moiety, which initiates a series of further transformations. The nature of the ring formed is determined by heteroatom bridging the two reactive units. The propargylamine derivative in **3.88.**, for example gave a pyrrole ring.[112]

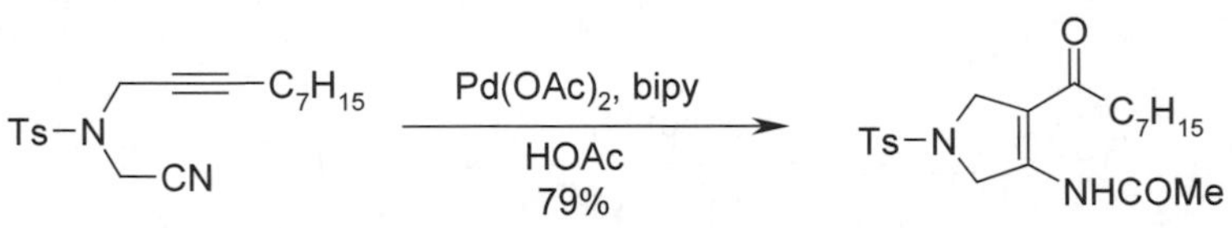

(3.88.)

In a formally similar reaction an alkene and an alkyne moiety were coupled in the presence of a palladium catalyst in formic acid. The reaction cascade in this case starts by the hydropalladation of the triple bond, which is followed by a Heck-type carbon-carbon bond formation. Running the process in the presence of a chiral catalyst the furane derivative was formed in acceptable enantiomeric excess (**3.89.**).[113]

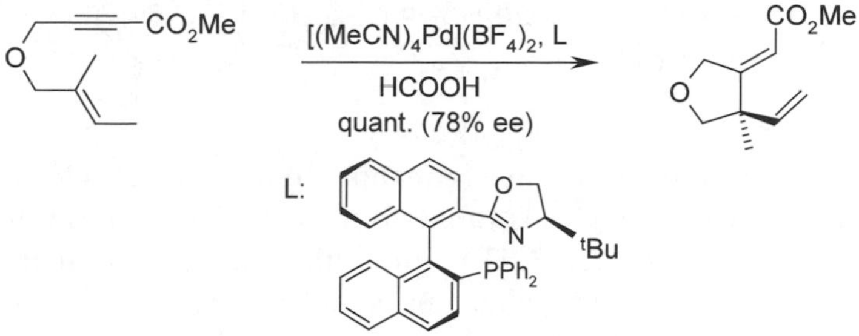

$$(3.89.)$$

Imines, ethyl acetylenedicarboxylate and benzoyl chloride were combined in the presence of carbon monoxide and a palladium-tri-*o*-tolylphosphine catalyst system to pyrrole derivatives (**3.90.**). Although the carbon monoxide is formally oxidized to carbon dioxide, during the catalytic cycle it is inserted into the intermediates formed and is extruded in a retro-Diels-Alder reaction only in the concluding step of the reaction sequence.[114]

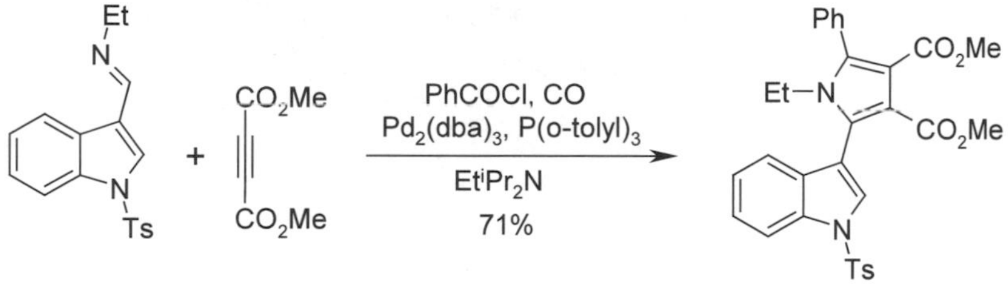

$$(3.90.)$$

Dipropargyl ethers can be cyclized to isobenzofurane derivatives both in palladium or nickel catalysed transformations. In the former case dipent-2-ynyl ether was coupled with allyl tosylate to give the corresponding bicycle, albeit in poor yield.[115] The same ring system was obtained in good yield in the nickel catalysed reductive cyclization of the diyne and the allene shown in **3.91**.[116]

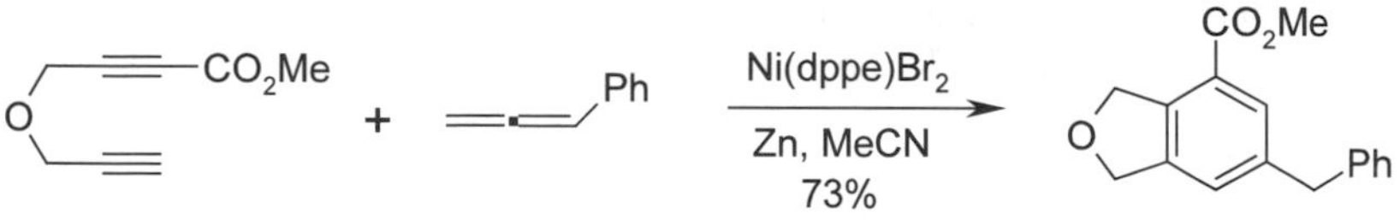

$$(3.91.)$$

Methylidenecyclopropanes were converted in a formal [3+2] cycloaddition both to furane and pyrrole derivatives. Their reaction, in the presence of a suitable palladium catalyst, with a carbonyl group led to the construction of a tetrahydrofurane ring (**3.92.**).[117] The analogous reaction in the presence of six membered heterocycles led to the formation of bridgehead nitrogen atom containing condensed systems.[118]

$$(3.92.)$$

The palladium catalysed transformation of *N*-propargyl-6-iodo-2-pyridones and isonitriles also leads to the formation of bridgehead nitrogen atom containing heterocycles (**3.93.**). From the mechanistic point it is interesting to note that the arylpalladium complex formed in the opening oxidative addition step undergoes 1,1-addition onto the isonitrile, thereby opening up the way for the following carbopalladation, C-H activation sequence.[119]

$$(3.93.)$$

3.5   REFERENCES

[1] Yue, W. S.; Li, J. J. *Org. Lett.* **2002**, *4*, 2201.

[2] Zhang, T. Y.; Zang, H. *Tetrahedron Lett.* **2002**, *43*, 193.

[3] Satoh, M.; Miyaura, N.; Suzuki, A. *Synthesis* **1987**, 373.

[4] Rutherford, J. L.; Rainka, M. P.; Buchwald, S. L. *J. Am. Chem. Soc.* **2002**, *124*, 15168.

[5] Timári, G.; Soós, T.; Hajós, Gy. *Synlett* **1997**, 1067.

[6] Csányi, D.; Timári G.; Hajós, Gy. *Synth. Commun.* **1999**, *22*, 3959.

[7] Béres, M.; Timári, G.; Hajós, Gy. *Tetrahedron Lett.* **2002**, *43*, 6035.

[8] Soós, T.; Timári, G.; Hajós, Gy. *Tetrahedron Lett.* **1999**, *40*, 8607.

[9] For a review of the pioneering works see Hegedus, S. L. *Angew. Chem. Int. Ed.* **1988**, *27*, 1113.

[10] Sundberg, R. J.; Pitts, W. J. *J. Org. Chem.* **1997**, *59*, 192.

[11] Li, J. J. *J. Org. Chem.* **1999**, *64*, 8425.

[12] Sakamoto, T.; Kondo, Y.; Uchiyama, M.; Yamanaka, H. *J. Chem. Soc., Perkin Trans. 1* **1993**, 1941.

[13] Evans, P.; Grigg, R.; Ramzan, M. I.; Sridharan, V.; York, M. *Tetrahderon Lett.* **1999**, *40*, 3021.

[14] Bosch, J.; Roca, T.; Armengol, M.; Fernández-Forner, D. *Tetrahedron* **2001**, *57*, 1041.

[15] Mori, M.; Chiba, K.; Ban, Y. *Tetrahedron Lett.* **1977**, *12*, 1037.

[16] Wensbo, D.; Annby, U.; Gronowitz, S. *Tetrahedron* **1995**, *51*, 10323.

[17] Mori, M.; Nakanishi, M.; Kajishima, D.; Sato, Y. *J. Am. Chem. Soc.* **2003**, *125*, 9801.

[18] Dounay, A. B.; Overman, L. E.*Chem. Rev.* **2003**, *103*, 2945.

[19] Dounay, A. B.; Hatanaka, K.; Kodanko, J. J.; Oestereich, M.; Overman, L. E.; Pfeifer, L. A.; Weiss, M. M. *J. Am. Chem. Soc.* **2003**, *125*, 6261.

[20] Oestereich, M.; Dennison, P. R.; Kodanko, J. J.; Overman, L. E. *Angew. Chem. Int. Ed.* **2001**, *40*, 1439.

[21] (a) Grigg, R.; Sansano, J. M.; Santhakumar, V.; Sridharan, V.; Thangavelanthum, R.; Thonrton-Pett, M.; Wilson, D. *Tetrahedron* **1997**, *53* 11803. (b) Grigg, R.; Sridharan, V.; Thang, J. *Tetrahedron Lett.* **1999**, *40*, 8277.

[22] Watanabe, T.; Arai, S.; Nishida, A. *Synlett* **2004**, *5*, 907.

[23] Chen, C-y.; Lieberman, D. R.; Larsen, R. D.; Verhoeven, T. R.; Reider, P. J. *J. Org. Chem.* **1997**, *62*, 2676.

[24] Yamazaki, K.; Nakamura, Y.; Kondo, Y. *J. Org. Chem.* **2003**, *68*, 6011.

[25] Pays, C.; Mangeney, P. *Tetrahedron Lett.* **2001**, *42*, 589.

[26] Grigg, R.; Sridharan, V.; Stevenson, P.; Sukirthalingam, S. *Tetrahedron* **1989**, *45*, 3557.

[27] Arnau, N.; Moreno-Manas, M.; Pleixats, R. *Tetrahedron* **1993**, *49*, 11019.

[28] Henke, B. R.; Aquino, C. J.; Birkemo, L. S.; Croom, D. K.; Dougherty, R. W. Jr.; Ervin, G. N.; Grizzle, M. K.; Hirst, G. C.; James, M. K.; Johnson, M. F.; Queen, K. L.; Sherrill, R. G.; Sugg, E. E.; Suh, E. M.; Szewczyk, J. W.; Unwalla, R. J.; Yingling, J.; Willson, T. M. *J. Med. Chem.* **1997**, *40*, 2706.

[29] Zhang, H.-C.; Maryanoff, B. E. *J. Org. Chem.* **1997**, *62*, 1804.

[30] Diaz, P.; Gendre, F.; Stella, L.; Charpentier, B. *Tetrahedron* **1998**, *54*, 4579.

[31] Trost, B. M.; Tang, W. *Angew. Chem. Int. Ed.* **2002**, *41*, 2795.

[32] (a) Burns, B.; Grigg, R.; Sridharan, V.; Stevenson, P.; Sukirthalingam, S.; Worakun, T. *Tetrahedron Lett.* **1989**, 30, 1135. (b) Luo, F.-T.; Wang, R.-T. *Heterocycles*, **1991**, *32*, 2365.

[33] Brown, D.; Grigg, R.; Sridharan, V.; Tambyrajah, V.; Thornton-Pett, M. *Tetrahedron* **1998**, *54*, 2595.

[34] Burns, B.; Grigg, R.; Sridharan, V.; Worakun, T. *Tetrahedron Lett.* **1988**, *29*, 4325.

[35] Witulski, B.; Alayrac, C.; Tevzadze-Saeftel, L. *Angew. Chem. Int. Ed.* **2003**, *42*, 4257.

[36] Olivero, S.; Clinet, J. C.; Dunach, E. *Tetrahedron Lett.* **1995**, *36*, 4429.

[37] Luo, F. T.; Wang, R. T. *Heterocycles* **1990**, *31*, 2181.

[38] Brown, D.; Grigg, R.; Sridharan, V.; Tambyrajah, V.; Thornton-Pett, M. *Tetrahedron* **1998**, *54*, 2595.

[39] Grigg, R.; Savic, V.; Sridharan, V.; Terrier, C. *Tetrahedron* **2002**, *58*, 8613.

[40] Arnau, N; Moreno-Mañas, M.; Pleixats, R. *Tetrahedron*, **1993**, *49*, 11019.

[41] Yang, C.-C.; Sun. P.-J.; Fang, J.-M. *J. Chem. Soc., Chem. Commun.* **1994**, 2629.

[42] Wu, X.; Mahalingam, A. K.; Wan, Y.; Alterman, M. *Tetrahedron Lett.* **2004**, *42*, 4635.

[43] Gabriele, B.; Veltri, L.; Salerno, G.; Costa, M.; Chiusoli, G. P. *Eur. J. Org. Chem.* **2003**, 1722.

[44] Kamitani, A.; Chatani, N.; Murai, S. *Angew. Chem.* **2003**, *115*, 1435.

[45] Wolfe, J. P.; Rennels, R.A.; Buchwald, S. L. *Tetrahedron* **1996**, *52*, 7525.

[46] Peat, A. J.; Buchwald, S. L. *J. Am. Chem. Soc.* **1996**, *118*, 1028.

[47] Cämmerer, S. S.; Viciu, M. S.; Stevens, E. D.; Nolan, S. P. *Synlett*, **2003**, *12*, 1871.

[48] Vincze, Z.; Bíró, A. B.; Csékei, M.; Kotschy, A. *unpublished results*.

[49] Bytschkov, I.; Siebeneicher, H.; Doye, S. *Eur. J. Org. Chem.* **2003**, 2888.

64 *Heterocycles from transition metal catalysis*

[50] Nozaki, K.; Takahashi, K.; Nakano, K.; Hiyama, T.; Tang, H.-Z.; Fujiki, M.; Yamaguchi, S.; Tamao, K. *Angew. Chem.* **2003**, *115*, 2097.

[51] Yamazaki, K.; Nakamura, Y.; Kondo, Y. *J. Chem. Soc., Perkin Trans. 1*, **2002**, 2137.

[52] Evindar, G.; Batey, R. A. *Org. Lett.* **2003**, *5*, 134.

[53] Kuwabe, S.-I.; Torraca, K. E.; Buchwald, S. L. *J. Am. Chem. Soc.* **2001**, *123*, 12202.

[54] Zhu, J.; Price, B. A.; Zhao, S. X.; Skonezny, P. M. *Tetrahedron Lett.* **2000**, *41*, 4011.

[55] Dai, W.-M.; Guo, D.-S.; Sun, L.-P. *Tetrahedron Lett.* **2001**, *42*, 5275.

[56] Wolf, L. B.; Tjen, K. C. M.; ten Brink, H. T.; Blaauw, R. H.; Hiemstra, H.; Schoemaker, H. E.; Rutjes, F. P. J. T. *Adv. Synth. Catal.* **2002**, *344*, 70.

[57] Utimoto, K.; Miwa, H.; Nozaki, H.; Tetrahedron *Lett.* **1981**, *22*, 4277.

[58] Arcadi, A.; Cacchi, S.; Marinelli, F. *Tetrahedron Lett.* **1989**, *30*, 2581.

[59] Rudisill, D. E.; Stille, J. K. *J. Org. Chem.* **1989**, *54*, 5856.

[60] Taylor, E. C.; Katz, A. H.; Salgado-Zamora, H.; McKillop, A. *Tetrahedron Lett.* **1985**, *26*, 5963.

[61] Tumkevicius, S.; Masevicius, V. *Synlett* **2004**, *13*, 2327.

[62] Dai, W.-M.; Guo, D.-S.; Sun, L.-P.; Huang, X.-H. *Org. Lett.* **2003**, *5*, 2919.

[63] Kabalka, G. W.; Wang, L.; Pagni, R. M. *Tetrahedron* **2001**, *57*, 8017.

[64] Cacchi, S.; Fabrizi, G. Paraisi, L. M. *Synthesis* **2004**, *11*, 1889.

[65] Kimball, D. B.; Weakley, T. J. R.; Haley, M. M. *J. Org. Chem.* **2002**, *67*, 6395.

[66] Seiller, B.; Bruneau, C.; Dixneuf, P. H. *Tetrahedron* **1995**, *51*, 13089.

[67] Qing, F.-L.; Gao, W.-Z.; Ying, J. *J. Org. Chem.* **2000**, *65*, 2003.

[68] Uozumi, Y.; Kobayashi, Y. *Heterocycles* **2003**, *59*, 71.

[69] Dai, W.-M.; Lai, K. W. *Tetrahedron Lett.* **2002**, *43*, 9377.

[70] Fancelli, D.; Fagnola, M. C.; Severino, D.; Bedeschi, A. *Tetrahedron Lett.* **1997**, *38*, 2311.

[71] Bates, C. G.; Saejueng, P.; Murphy, J. M.; Venkataraman, D. *Organic Lett.* **2002**, *4*, 4727.

[72] Cacchi, S.; Fabrizi, G.; Goggiomani, A. *Heterocycles* **2002**, *56*, 613.

[73] Houpis, I. N.; Choi, W. B.; Reider, P. J.; Molina, A.; Churchill, H.; Lynch, J.; Volante, R. P. *Tetrahedron Lett.* **1994**, *35*, 9355.

[74] McGuigan, C.; Yarnold, C. J.; Jones, G.; Velézquez, S.; Barucki, H.; Brancale, A.; Andrei, G.; Snoeck, R.; DeClercq, E.; Balzarin, J. *J. Med. Chem.* **1999**, *43*, 4479.

[75] Novák, Z.; Timári, G.; Kotschy, A. *Tetrahedron* **2003**, *59*, 7509.

[76] Novák, Z.; Csékei, M.; Timári, G.; Kotschy, A. *Arkivoc*, **2004**, 285.

[77] Hu, Y.; Nawoschik, K. J.; Liao, Y.; Ma, J.; Fathi, R.; Yang, Z. *J. Org. Chem.* **2004**, *69*, 2235.

[78] Hu, Y.; Zhang, Y.; Yang, Z.; Fathi, R. *J. Org. Chem.* **2002**, *67*, 2365.

[79] Yue, D.; Larock, R. C. *J. Org. Chem.* **2002**, *67*, 1905.

[80] Genet, J. P.; Balabane, M.; Bäckwall, J. E.; NyStöm, J. E. *Tetrahedron Lett.* **1983**, *24*, 2745.

[81] Macsári, I.; Szabó, K. *Chem. Eur. J.* **2001**, *7*, 4097.

[82] Anderson, P. G.; Bäckwall, J.-E.; *J. Am. Chem. Soc.* **1992**, *114*, 8696.

[83] Harrington, P. J.; Hegedus, L. S.; McDaniel, K. F. *J. Am. Chem. Soc.* **1987**, *109*, 4335 and references therein.

[84] Hegedus, L. S.; Allen, G. F.; Bozell, J. J.; Waterman, E. L. *J. Am. Chem. Soc.* **1978**, *100*, 5800.

[85] Hegedus, L. S.; Allen, G. F.; Olsen, D. J. *J. Am. Chem. Soc.* **1980**, *102*, 3582.

[86] Krishniuda, K.; Krishna, P. R.; Mereyala, H. B. *Tetrahedron Lett.* **1996**, *37*, 6007.

[87] Shibata, T.; Kadowaki, S.; Takagi, K. *Heterocycles*, **2002**, *57*, 2261.

[88] Zenner, J. M.; Larock, R. C. *J. Org. Chem.* **1999**, *64*, 7312 and references therein.

[89] Lutete, L. M.; Kadota, I.; Shibuya, A.; Yamamoto, Y. *Heterocycles* **2002**, *58*, 347.

[90] Ma, S.; Zhang, J.; Lu, L. *Chem. Eur. J.* **2003**, *9*, 2447.

[91] Ma, S. Yu, Z. *Angew. Chem. Int. Ed.* **2002**, *41*, 1775.

[92] Kel'in, A. V.; Gevorgyan, V. *J. Org. Chem.* **2002**, *67*, 95.

[93] Kim, J. T.; Kel'in, A. V.; Gevorgyan, V. *Angew. Chem.* **2003**, *115*, 102.

[94] (a) Tollari, S.; Cenini, S.; Crotti.; Gianella, E. *J. Mol. Cat.* **1994**, *87*, 203. (b) Tollari, S.; Cenini, S.; Rossi, A.; Palmisano, G. *J. Mol. Cat.* **1998**, *135*, 241.

[95] Smitrovich, J. H.; Davies, I. W. *Org. Lett.* **2003**, *6*, 534.

[96] Itahara, T.; Sakakibara, T. *Synthesis* **1978**, 607.

[97] Black, D. St C.; Kumar, N.; Wong, L. C. H. *J. Chem. Soc. Chem. Commun.* **1985**, 1174.

[98] Knölker, H.-J.; Reddy, K. R. *Heterocycles* **2003**, *60*, 1049.

[99] Liu, Z.; Larock, R. C. *Org. Lett.* **2004**, *6*, 3739.

[100] Jonckers, T. H. M.; Maes, B. U. W.; Lemière, G. L. F.; Rombouts, G.; Pieters, L.; Haemers, A.; Dommisse, R. A. *Synlett* **2003**, *5*, 615.

[101] Harris, J. M.; Padwa, A. *Org. Lett.* **2003**, *5*, 4195.

[102] Dajka-Halász, B.; Monsieurs, K.; Éliás, O.; Károlyházy. L.; Tapolcsányi, P.; Maes, B. U. W.; Riedl, Zs.; Hajós, Gy.; Dommisse, R. A.; Lemière, G. L. F.; Košmrlj, J.; Mátyus, P. *Tetrahedron* **2004**, *60*, 2283.

[103] Harayama, T.;Morikami, Y.; Shigeta, Y.; Abe, H.; Takeuchi, Y. *Synlett* **2003**, *6*, 847.

[104] Larock, R. C.; Doty, M. J.; Han, X. *Tetrahedron Lett.* **1998**, *39*, 5143.

[105] (a) Larock, R. C.; Yum, E. K. *J. Am. Chem. Soc.* **1991**, *113*, 6689. (b) Larock, R. C.; Yum, E. K.; Refvik, M. D. *J. Org. Chem.* **1998**, *63*, 7652

[106] Smith, A. L.; Stevenson, G. I.; Swain, C. J.; Castro, J. L. *Tetrahedron Lett.* **1998**, *39*, 8317.

[107] (a) Larock, R. C.; Yum, E. K.; Doty, M. J.; Sham, K. K. C. *J. Org. Chem.* **1995**, *60*, 3270. (b) Bishop, B. C.; Cottrell, I. F.; Hands, D. *Synthesis* **1997**, 1315.

[108] Wensbo, D.; Gronowitz, S. *Tetrahedron* **1996**, *52*, 14975.

[109] (a) Park, S. S.; Choi, J.-K.; Yum, E. K.; Ha, D.-C. *Tetrahedron Lett.* **1998**, *39*, 627-630. (b) Kang, S. K.; Park, S. S.; Kim, S. S.; Choi, J.-K.; Yum, E. K. *Tetrahedron Lett.* **1999**, *40*, 4379.

[110] Larock, R. C.; Berrios-Pena, N.; Narayanan, K. *J. Org. Chem.* **1990**, *55*, 3447.

[111] Rozhkov, R. V.; Larock, R. C. *J. Org. Chem.* **2003**, *68*, 6314.

[112] 102 Zhao, L.; Lu, X. *Angew. Chem. Int. Ed.* **2002**, *41*, 4343.

[113] Hatano, M.; Yamanaka, M.; Mikami, K. Eur. *J. Org. Chem.* **2003**, *68*, 2552.

[114] Dhawan, R.; Arndsten, B. A. *J. Am. Chem. Soc.* **2004**, *126*, 469.

[115] Tsukada, N.; Sugawara, S.; Nakaoka, K.; Inoue, Y. *J. Org. Chem.* **2003**, *68*, 5961.

[116] Shanmugasundaram, M.; Wu, M.-S.; Jeganmohan, M.; Huang, C.-W.; Cheng, C.-H. *J. Org. Chem.* **2002**, *67*, 7724.

[117] Nakamura, I.; Oh, B. H.; Saito, S.; Yamamoto, Y. *Angew. Chem. Int. Ed.* **2001**, *40*, 1298.

[118] Siriwardana, A. I.; Nakamura, I.; Yamamoto, Y. *J. Org. Chem.* **2004**, *69*, 3202.

[119] Curran, D. P.; Du, W. *Organic Lett.* **2002**, *4*, 3215.

Chapter 4

THE SYNTHESIS OF SIX MEMBERED RINGS

The transition metal catalyzed synthesis of six membered heterocycles attracted less attention than the preparation of five membered rings. The majority of examples discussed in this chapter achieves the formation of the ring through the combination of two fragments, forming a carbon-carbon and a carbon-heteroatom bond.

4.1   TRANSMETALATION ROUTE

Examples of the formation of six membered heterocycles through cross-coupling reactions are rare. Although several procedures rely on the use of coupling reactions for the connection of two fragments through a carbon-carbon bond, the final ring closing step is usually the formation of a carbon-heteroatom bond through addition or substitution. These reactions will be discussed in Chapter 4.3. As the presented examples will underline, the formation of the parent heterocycles is not viable this way, and the described procedures were developed for the preparation of condensed systems.

An excellent example of the formation of a six membered ring is the hexabutyldistannane mediated ring closure of the brominated indole derivative shown in **4.1**. The first step of the process is the palladium catalyzed exchange of one of the bromines to a tributylstannyl moiety, followed by the closure of the six membered ring in a Stille coupling.[1]

67

(4.1.)

The examples presented in **4.2.-4.7.** describe the preparation of condensed nitrogen containing ring systems, where the connection of the fragments is achieved in a similar manner. Formation of the carbon-carbon bonds by Suzuki coupling is accompanied by the formation of the carbon-nitrogen bond using classical organic transformations (Schiff-base formation, electrophilic substitution etc.). Although the reactions where carbon-carbon bond formation precedes formation of the carbon-nitrogen bond formally fall outside the scope of this chapter, we mention some examples due to the synthetic importance of this approach (*N.B.* the same approach has also been utilised in the preparation of coumarin derivatives[2]).

Thieno-pyridopyridine isomers were prepared by the coupling of a halogenated aminopyridine derivative and formyl-thienylboronic acid. The coupling of the unprotected reagents led to the concomitant formation of the carbon-nitrogen and carbon-carbon bonds in good yield (**4.2.**).[3]

(4.2.)

The coupling of the analogous *N*-oxide required the protection of the amino group. Suzuki coupling under the same conditions as in **4.2.**, followed by the *in situ* removal of the acyl protecting group and spontaneous ring closure led to the desired tricyclic *N*-oxide in good yield (**4.3.**).[4]

(4.3.)

Condensation of two thiophenes with a pyridine ring, leading to dithienopyridine, requires the coupling of a formylthiophene and an aminothiophene derivative. In this case both functional groups had to be protected prior to the coupling. The use of acid sensitive protecting groups allowed for the parallel deprotection of both functional groups after the cross-coupling (Stille coupling), and the subsequent formation of the central heterocycle (**4.4.**).[5] The mediocre yield of the process was attributed to the steric bulk of the Boc-group on the halothiophene, hindering the transmetalation step.

N-protected *ortho*-anilineboronic acid was used by Mátyus and co-workers to introduce an annelating cinnoline ring onto pyrimidine derivatives. The connection of the phenyl and pyrimidine moieties by Suzuki coupling was followed by the establishment of the carbon-nitrogen bond through a diazo-coupling (**4.5.**).[6]

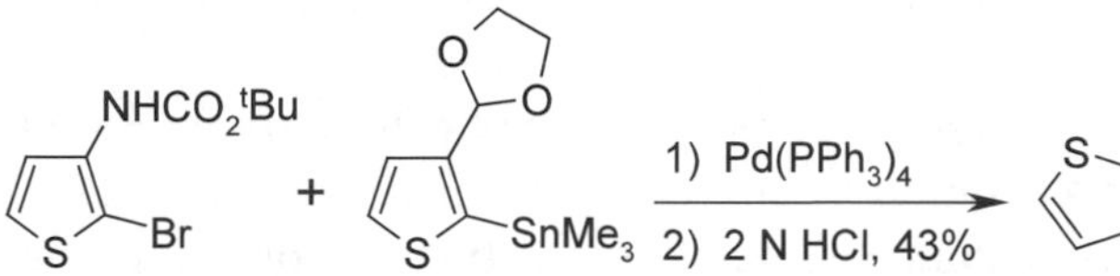

(4.4.)

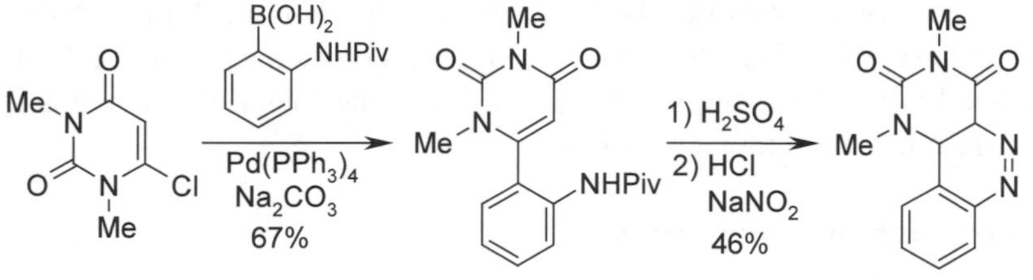

(4.5.)

The intramolecular variant of the palladium catalyzed coupling of aryl halides with classical carbanoins, sometimes considered a variant of the Buchwald-Hartwig coupling, might also be used for the formation of heterocyclic systems. 4-Amino-cyclohexanone derivatives, bearing a pendant aryl or vinyl halide moiety undergo ring closure in the presence of a suitable base and catalyst. The cyclization of *N*-(2-bromocrotyl)-4-aminocyclohexanone in the presence of a palladium-triphenylphosphine catalyst system and potassium *tert*-butoxide led to the formation of the 2-azabicyclo[3.3.1]nonane system in good yield (**4.6.**).[7]

Starting from the analogous *N*-(2'-iodophenyl)-4-aminocyclohexanone derivative the bridged benzazocine, a framework present in several natural products, was isolated in near quantitative yield (**4.7.**).[8]

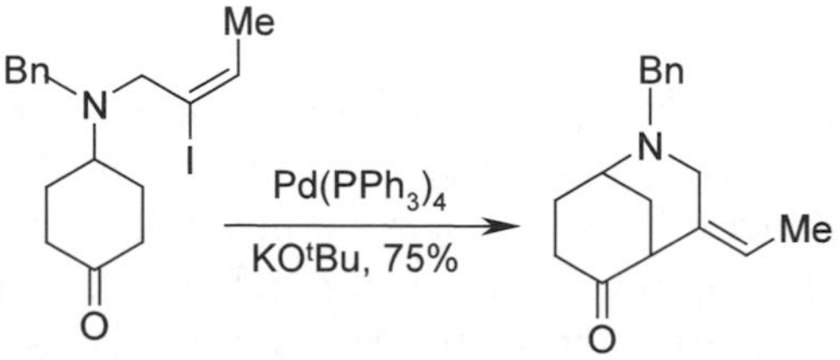

(4.6.)

$$(4.7.)$$

4.2   INSERTION ROUTE

Transition metal catalyzed insertion reactions offer a variety of alternate approaches for the preparation of six membered heterocyclic rings. Besides intramolecular Heck-couplings, annulation reactions involving the insertion of an acetylene derivative, in most cases accompanied by the incorporation of carbon monoxide constitute the majority of this chapter. Although some of these processes involve the formation of a carbon-heteroatom bond, they are discussed here, while the similar annulation reactions not involving CO insertion are mentioned in Chapter 4.4.

Heck reaction (olefin insertion)

The reaction of *N*-pentenyl-2-iodoindole in the presence of a palladium-triphenylphosphine catalyst led to the formation of a mixture of isomeric products in good yield (**4.8.**). Addition of thallium(I) acetate favoured the formation of an exocyclic double bond, while in its absence the product containing the endocyclic olefin moiety is formed preferentially. The shortening of the *N*-alkenyl chain by one carbon leads to the selective formation of a five membered ring.[9] Starting from indole-carboxamide derivatives both *β*-, and *γ*-carbolinones are available in intramolecular Heck coupling.

$$(4.8.)$$

Since the six membered rings formed by Heck coupling are usually partially saturated, introduction of chirality into the formed ring is possible through the proper choice of a chiral starting material. Chiral isoquinoline derivatives were prepared, for example, through the coupling of *ortho*-iodobenzoic acid derivatives, connecting the olefin moiety and the aromatic ring through an amino acid derived chiral linker. The chiral information is retained during the Heck reaction (**4.9.**). Addition of silver(I) nitrate to the catalyst system led to the selective dimerisation of the aryl moieties instead of the Heck reaction.[10]

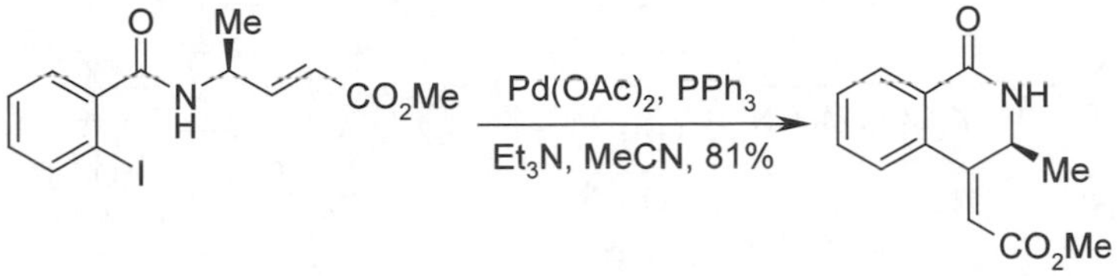

(4.9.)

N-substituted methylenephthalimide derivatives served as the starting material for the preparation of a series of tetracyclic systems. The intramolecular Heck reaction of these compounds (**4.10.**) led in each case to the formation of a carbon-carbon bond between the exocyclic carbon atom of the olefin bond and the aryl moiety. The formation of five, six and seven membered rings was achieved with near equal efficiency, enabling the preparation of some isoindolobenzazepine alkaloids.[11]

The natural product, anhydrolycorinone was prepared in the palladium mediated intramolecular Heck reaction of a tetrahydroindole derivative. The coupling, which is run in air and requires the use of a stoichiometric amount of palladium, is accompanied by the dehydrogenation of the cyclohexadiene moiety and the oxidation at the benzylic position to a lactam (**4.11.**).[12]

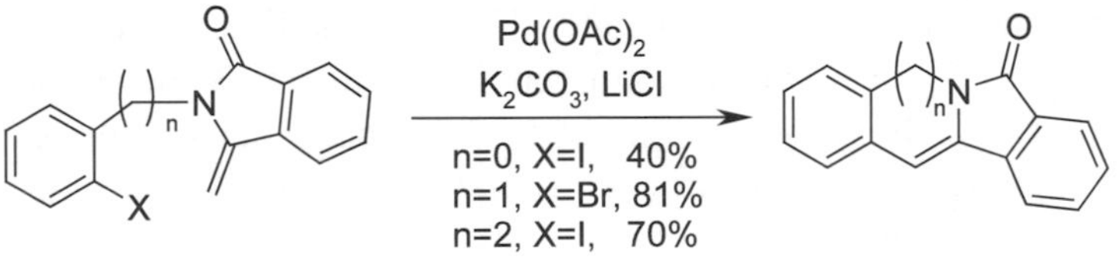

n=0, X=I, 40%
n=1, X=Br, 81%
n=2, X=I, 70%

(4.10.)

Pd(PPh$_3$)$_4$

DMF, air, 29%

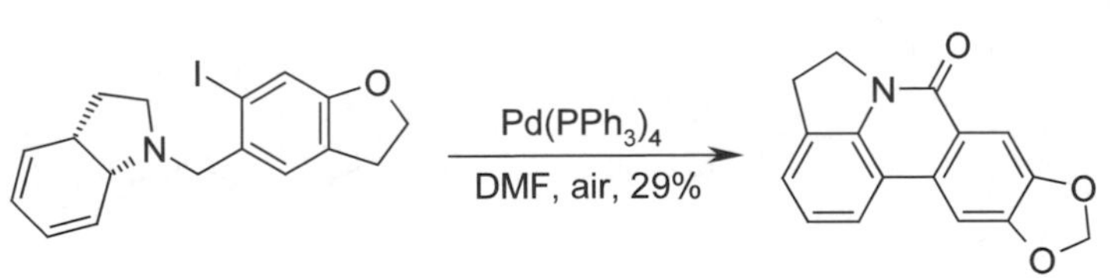

(4.11.)

The Heck coupling is also efficient in the formation of pyrane derivatives. By starting from the appropriately substituted 1,3,5-tribromobenzene derivative, three fused pyrane rings were constructed in the same reaction. The triple Heck reaction in the presence of palladium acetate and triphenylphosphine led to the formation of the tetracyclic product in good yield accompanied by a minor amount of the tricyclic intermediate(**4.12.**).[13]

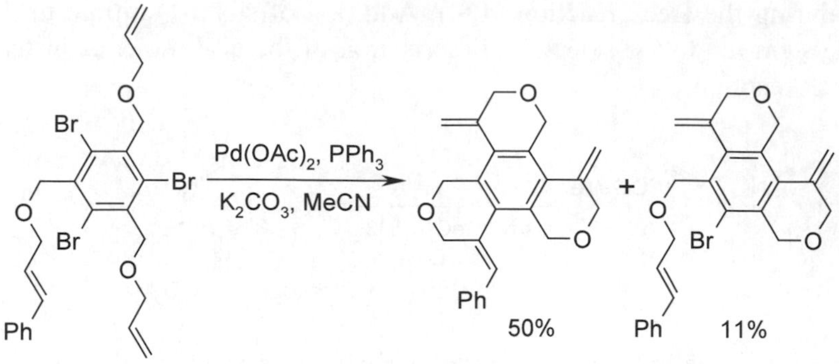

$$(4.12.)$$

Annulation through acetylene insertion

The insertion of acetylene derivatives might also be utilised in the preparation of six membered rings. A characteristic distinction between such processes and olefin insertion is the fact, that the intermediate formed by the insertion of an acetylene into the palladium-carbon bond is unable to undergo β-hydride elimination, therefore the concluding step of these processes is usually reductive elimination.

N-substituted 2-iodoindole-3-carboxylic acid derivatives incorporating an alkyne containing tether (*c.f.* **4.8.**) underwent ring closure in the presence of a catalytic amount of palladium acetate. The intermediate, formed in the insertion step, in this case attacks the phenyl substituent of the indole ring (C-H activation) and leads to the formation of another six membered ring (**4.13.**). The approach was also extended to the formation of an azepine ring using an elongated tether.[14]

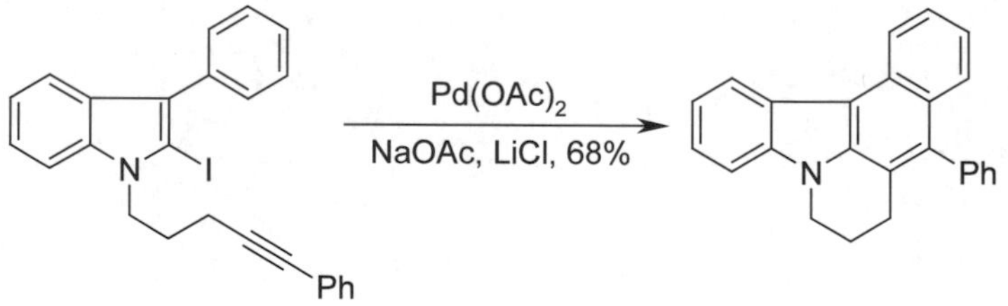

$$(4.13.)$$

The insertion of alkynes into arylpalladium complexes might also be accompanied by the insertion of carbon monoxide into the resulting vinylpalladium intermediate. The carbonylative annulation of *N*-protected 2-iodoanilines and internal alkynes under an ambient pressure of carbon monoxide resulted in the formation of 2-quinolones (**4.14.**). The protection of the nitrogen atom in the aniline is crucial to the success of the reaction.

While the use of acyl and sulfonyl protecting groups led to mediocre yield, masking of the nitrogen in the form of a carbamate led to improved yields.[15]

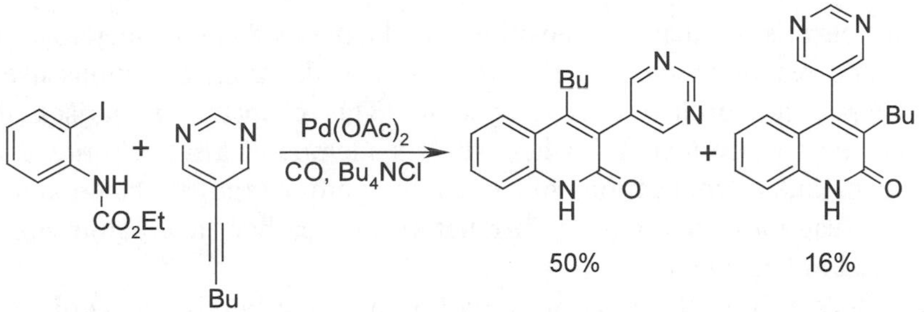

(4.14.)

The carbonylative annulation of *o*-halophenols can be exploited both in the synthesis of flavones and coumarins. The palladium catalyzed three component coupling of terminal acetylenes with 2-iodophenol and carbon monoxide usually leads to the formation of a mixture of flavones and aurones.[16] Starting from 2-acetoxy-iodobenzene instead of the phenol the formation of flavones proceeded with high selectivity and the spontaneous removal of the acetyl group under the applied conditions led to the desired product in high yield (**4.15.**).[17] In these processes the insertion of carbon monoxide precedes the attachment of the terminal acetylene, and it is the carbonylative coupling product that undergoes spontaneous ring closure.

The analogous palladium catalyzed reaction of internal acetylenes, 2-iodophenol and carbon monoxide leads to the selective formation of coumarins. The heterocyclic analogues of *o*-iodophenol are also effective. The *o*-iodopyridone shown in **4.16.** for example gave rise to the formation of azacoumarin in 70% yield.[18] In these processes the insertion of the acetylene derivative occurs in advance of the insertion of CO. Interestingly, the change of the acetylene to an alkene reverses the insertion order and leads to flavone formation.[19]

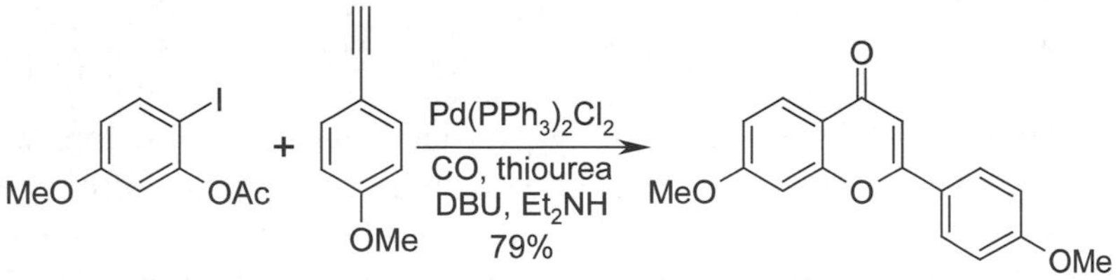

(4.15.)

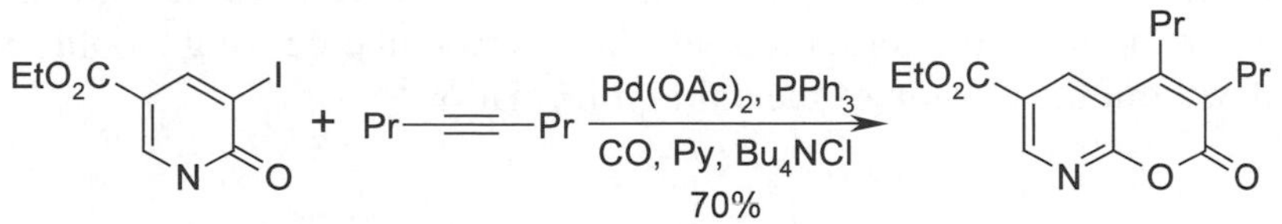

(4.16.)

4.3   CARBON-HETEROATOM BOND FORMATION

It is not surprising, that the transition metal catalyzed carbon-heteroatom bond forming reactions constitute a major part of the transformations used for the preparation of heterocyclic systems. This chapter summarises all those catalytic transformations, where the six membered ring is formed in the intramolecular connection of a carbon atom and a heteroatom. Annulation reactions, involving the formation of a carbon-heteroatom bond are discussed in Chapter 4.4.

The copper catalyzed formation of carbon-nitrogen bonds was exploited in the conversion of 2-(3'-aminopropyl)-bromobenzene to tetrahydro-isoquinoline (**4.17.**). The coupling, which proceeded readily already at 40 °C in the presence of diethyl salicylamide, was also efficient in the formation of dihydroindole derivatives.[20]

One of the rare examples, where the palladium catalyzed closure of a six membered ring leads to an aromatic heterocycle is presented in **4.18**. Intramolecular transformation of the 2-bromoindole derivative in the presence of a palladium-BINAP catalyst led to the formation of the α-carboline (pyrido[2,3-*b*]indole) skeleton. The ring closure was accompanied by the oxidation of the intermediate dihydrocarboline derivative.[21]

(4.17.)

(4.18.)

The palladium catalyzed carbon-oxygen bond forming reaction was utilised by Buchwald and co-workers in the synthesis of a series of oxygen containing heterocycles including partially reduced benzofurans, chromans, benzoxepins, benzodioxanes, and benzoxazepines. Starting from a chiral *N*-(*o*-bromophenyl)-β-aminoalcohol (**4.19.**), by the careful choice of the coupling conditions, the formation of the six membered ring could be achieved without the erosion of the chiral information.[22]

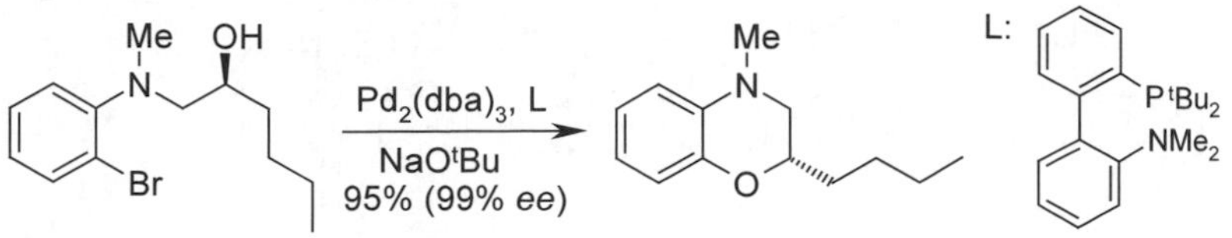

(4.19.)

The formation of chromane derivatives has also been realised in the palladium catalyzed intramolecular nucleophilic substitution of allyl carbonates (Tsuji-Trost reaction). In most cases the reaction is accompanied by the formation of a new centre of chirality. Using Trost's chiral ligand the ring closure was carried out in an enantioselective manner. The asymmetric allylation of the phenol derivative shown in **4.20.** was achieved both in good yield and with excellent selectivity.[23]

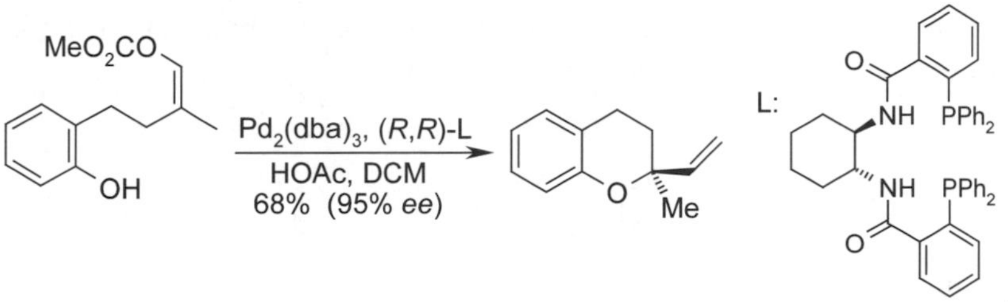

(4.20.)

Allylic diacetates were converted into *N,N*, *O,O*, and *N,O*-heterocycles in a palladium catalyzed two step nucleophilic displacement sequence. The reaction of *o*-aminophenol and cyclopentenyl-diacetate, for example, gave the cyclopentene condensed phenoxazine derivative in acceptable yield (**4.21.**), while *o*-phenylenediamine and catechol gave the corresponding phenazine and dioxine derivative in good yield.[24]

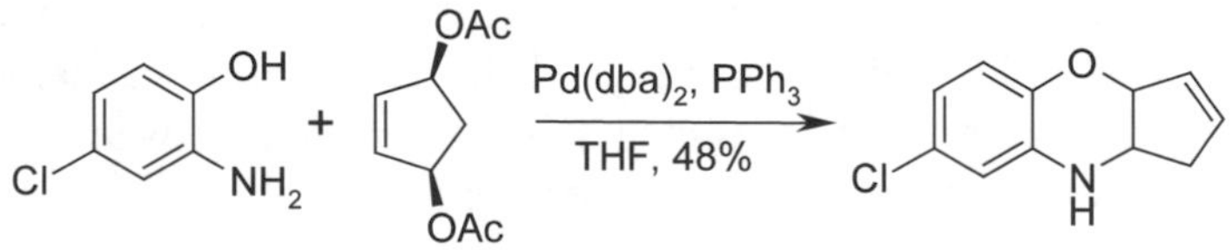

(4.21.)

Palladium was also effective in promoting the conversion of the indolyl-triflate shown in **4.22.** to the tricyclic indolopyrazine system. The reaction, proceeding by the incorporation of a molecule of ethylenediamine, also proceeded in the absence of the palladium catalyst, but it was more sluggish and gave lower yields. Tetracyclic indolo[2,3-*b*]quinoxaline was also prepared in the same manner, using *o*-phenylenediamine as nucleophile.[25]

$$(4.22.)$$

The intramolecular nucleophilic attack of a nitrogen atom on an allylpalladium complex was also used to construct a five and a six membered heterocycle in the same step. *N*-substituted 2-iodobenzamides bearing an allene function in the appropriate distance from the iodine underwent cyclization through the carbopalladation of the allene moiety by the arylpalladium complex, formed in the first step of the catalytic cycle. The intermediate allylpalladium complex, part of a nine membered ring, cyclized readily to give the pyrroloisoquinolone derivative in excellent yield (**4.23.**). The nature of the added ligand and the solvent both had a marked influence on the efficiency of the transformation.[26]

$$(4.23.)$$

Internal alkynes bearing a pendant sulfoxide group were found to undergo *syn*-selective palladium catalyzed sulfinylzincation. The opening diethylzinc mediated dealkynylation of the *tert*-butylethynyl-sulfoxides (**4.24.**) was followed by spontaneous ring closure, which led to the formation of a series of different sulphur containing heterocycles. The synthesis of five, six and seven membered rings proceeded equally well.[27] Although examples are sparse, the palladium catalysed formation of a six membered ring using intramolecular carbon-sulphur bond formation between an aryl iodide and a mercaptane moiety has also been reported.[28]

X: nil - 66%, CH$_2$ - 92%, C$_2$H$_4$ - 49%, O - 50%

$$(4.24.)$$

Starting from *o*-ethynylphenyl-urea derivatives the quinoline, quinazoline and benzoxazine ring systems could be prepared selectively (**4.25.**). The opening step of the process in each case is the desilylation of the acetylene, which is followed by the palladium(II) catalyzed attack of the nitrogen or oxygen atom of the urea moiety on the triple bond. The formation of the quinazoline and benzoxazine rings was rationalised by assuming a subsequent methoxycarbonylation step, while rearrangement of the primary benzoxazine product and subsequent oxidative methoxycarbonylation might be responsible for the formation of quinolines from the monosubstituted arylurea derivative.[29]

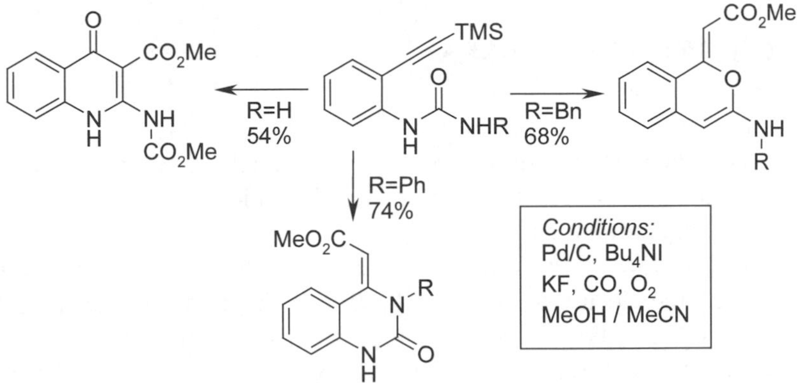

(4.25.)

In the presence of copper(I) salts in acidic media *o*-ethynyl-benzaldehyde derivatives were found to cycloisomerise to 2-benzopyrylium salts (**4.26.**). The reaction, although working in the absence of catalyst too, was accelerated by the addition of different metal salts. The reaction was applied in the preparation of azaphilones and related molecules.[30]

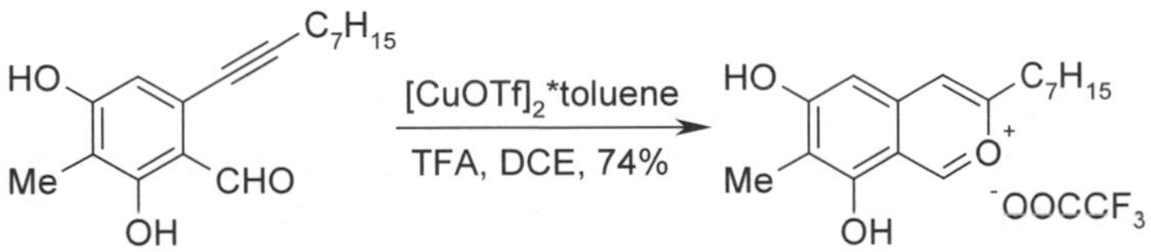

(4.26.)

Yamamoto reported an analogous transformation recently, where the copper catalyzed intramolecular cyclization of an *o*-ethynylbenzaldehyde derivative in the presence of an alcohol gave dihydro-benzopyrane in good yield (**4.27.**). All the copper(I) salts tested led to the selective formation of benzopyranes, copper(I) iodide being the most efficient, while copper(II) chloride initiated the formation of a 1:1 mixture of the desired benzopyrane and the isomeric benzofurane derivative. It is not clear yet, whether the ring closure precedes the alcohol addition and the reaction goes through a benzopyrylium intermediate, or the hemiacetal, formed by the addition of the alcohol, is the key intermediate in the copper catalyzed ring closure.[31]

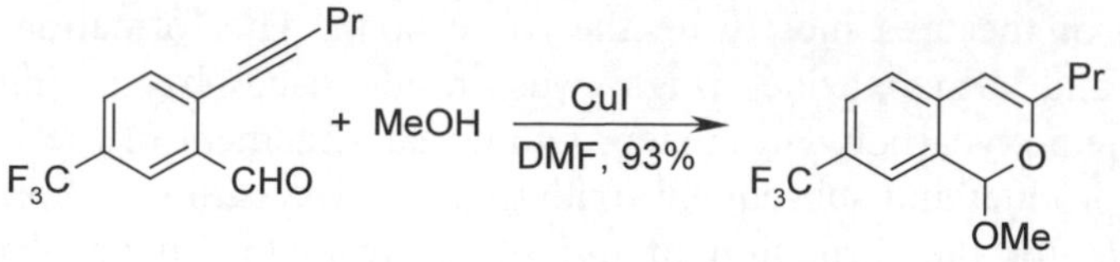

(4.27.)

4.4   OTHER PROCESSES

Besides ring closure reactions proceeding through C-H activation, a major part of this chapter is devoted to the formation of six membered heterocycles in annulation reactions.

Formation of the antiasthmatic imidazoquinoline compound was achieved through the closure of the central pyridine ring in a heteroaryl Heck reaction (**4.28.**). The best results were obtained in the presence of tetrabutylammonium chloride without any added ligand (Jeffery's "ligand free" variant).[32]

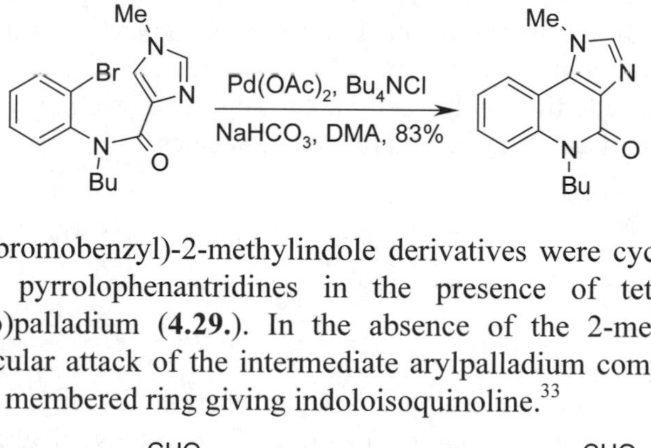

(4.28.)

N-(2'-bromobenzyl)-2-methylindole derivatives were cyclized relatively easily to pyrrolophenantridines in the presence of tetrakis(triphenyl-phosphino)palladium (**4.29.**). In the absence of the 2-methyl group the intramolecular attack of the intermediate arylpalladium complex is directed at the five membered ring giving indoloisoquinoline.[33]

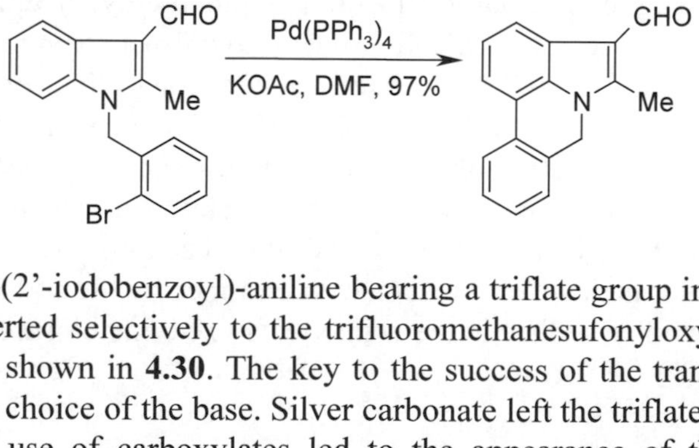

(4.29.)

The *N*-(2'-iodobenzoyl)-aniline bearing a triflate group in the 6'-position was converted selectively to the trifluoromethanesufonyloxy-phenantridone derivative shown in **4.30**. The key to the success of the transformation was the proper choice of the base. Silver carbonate left the triflate function intact, while the use of carboxylates led to the appearance of the carbonyloxy function in the product. Tertiary amines initiated the reduction of the triflate, and the hydrolysis byproduct was also present in both latter cases.[34] The same methodology was successfully extended to the preparation of partially reduced phenantridone derivatives,[35] pyrrolophenantridons,[36] and naphthobenzazepines.[37]

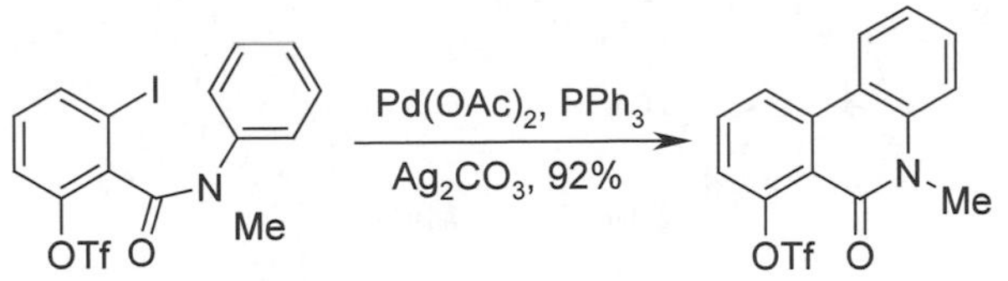

(4.30.)

The central pyridazine ring of the condensed indole derivative in **4.31.** was built up by the electrophilic attack of the pyridylpalladium intermediate on the indole ring.[38] By blocking the 2-position of indole through substitution the ring closure was directed into the *peri*-position forming a diazepine ring.[39]

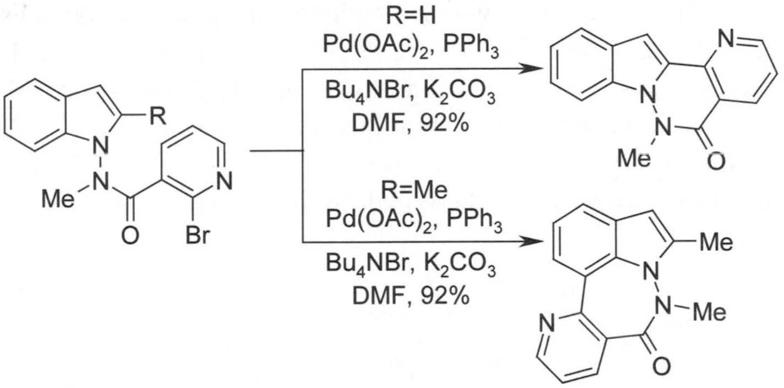

(4.31.)

The susceptibility of the indole ring towards electrophilic attack has also been exploited by Mérour in the annulation of a coumarin unit to the indole ring. The heating of the *o*-bromophenyl ester of indole-2-carboxylic acid in the presence of a palladium-triphenylphosphine catalyst led to the formation of the tetracyclic product in 66% yield (**4.32.**).[40]

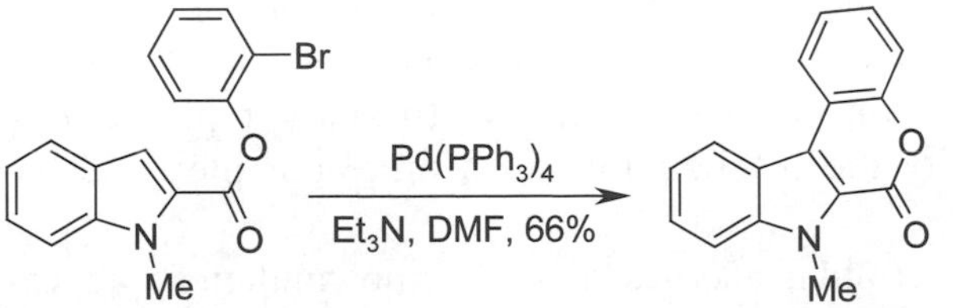

(4.32.)

In certain cases the oxidative coupling of two aromatic rings can be achieved in a double C-H activation. Pyrrolophenantridones were prepared by the palladium acetate mediated closure of the six membered central fragment (**4.33.**). The setback of the procedure, besides the mediocre yield is the necessity to use a stoichiometric amount of palladium.[41]

(4.33.)

The palladium catalyzed iminoannulation and carboxyannulation of alkynes and an appropriate aryl/vinyl halide is an efficient tool to construct six membered nitrogen and oxygen heterocycles. The process encompasses the concomitant formation of a carbon-carbon and a carbon-heteroatom bond.

Most annulation procedures, which produce nitrogen heterocycles start from a *tert*-butylimine, where the sacrificial organic moiety is released in the course of the process as isobutene. Ring closure of the 2-iodoveratraldehyde derivative shown in **4.34.** and ethyl phenylpropiolate, for example, led to the formation of the appropriate isoquinoline derivative in excellent yield. Formation of the isomeric 4-phenylisoquinoline compound was also observed (5%).[42]

(4.34.)

The analogous transformation of 2-formyl-3-iodoindole derivatives and phenylacetylene gave *β*-carbolines in good yield (**4.35.**). The efficiency of the procedure was moderately influenced by the nature of indole's *N*-substituent.[43]

The formation of phenantridone analogues from *o*-halobenzamides and haloarenes, reported by Catellani, could be considered a vaiation of this approach, where the acetylene bond is formally replaced by a benzyne moiety. In reality the reaction follows a more complex and mechanisticly distinct pathway.[44]

The use of terminal alkynes in the iminoannulation process follows a different mechanistic pathway. The coupling of *N*-methyl-3-formyl-2-iodoindole and phenylacetyle was carried out in a stepwise manner, consisting of the Sonogashira coupling of the haloindole and the acetylene, the formation of the *tert*-butylimine and finally the copper catalyzed closure of the six membered ring (**4.36.**), all steps giving a near quantitative yield.[45,46] Ring closure of the intermediate alkynylimine can also be initiated by the addition of an electrophile, which also appears in the formed six membered ring.[47]

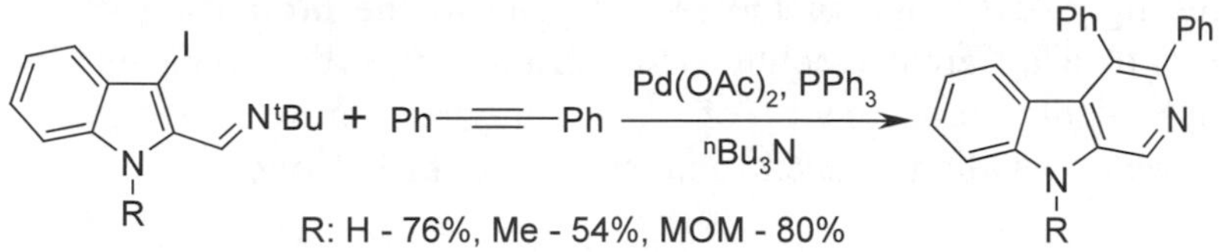

$$\text{R: H - 76\%, Me - 54\%, MOM - 80\%}$$

(4.35.)

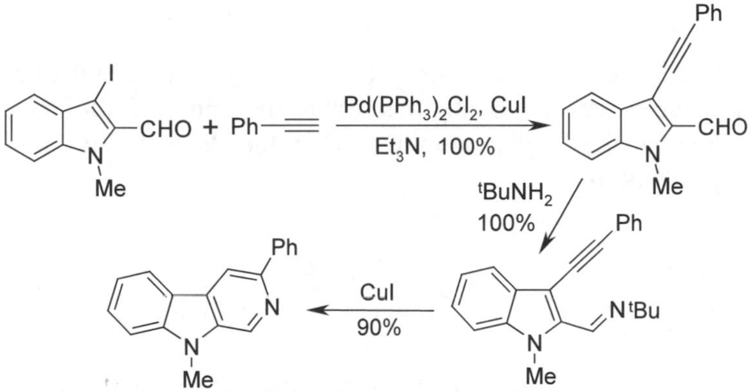

(4.36.)

In spite of its formal similarity to the above mentioned annulation processes, the reaction shown in **4.37.** includes a unique migration step. Oxidative insertion of the palladium into the phenyl-iodine bond is followed by the migration of the palladium onto the more electron rich indole ring. The 2-indolylpalladium complex than carbopalladates the pendant alkyne moiety and the process ends by the formal activation of a C-H bond of the phenyl substituent and subsequent reductive elimination, furnishing the pentacyclic product.[48] The same strategy has been utilised in the preparation of the indoloindolone framework from *N*-benzoyl-3-(o-iodophenyl)-indole in an oxidative addition – palladium migration – C-H activation sequence.[49]

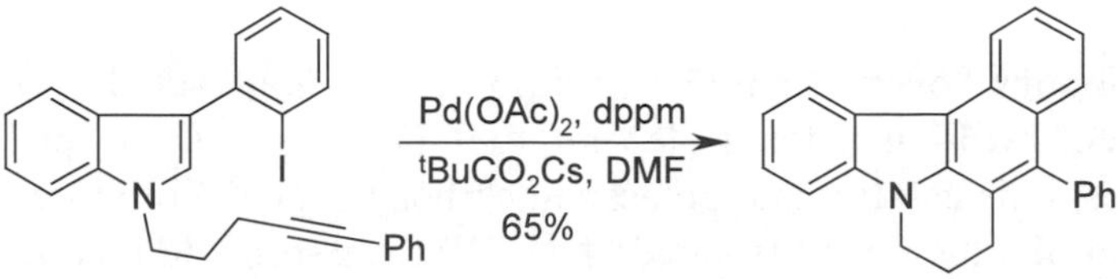

(4.37.)

The annulation reactions, which start from carboxylic acids and esters usually lead to the formation of pyrone derivatives. Methyl (Z)-3-iodoacrylate and 3-hexyne gave, for example, 5,6-diethyl-2-pyrone in acceptable yield (**4.38.**). Inclusion of the acrylate into a six membered ring starting from ethyl 2-bromocyclohexen-1-carboxylate, led to a condensed ring system, giving a partially reduced isocoumarin derivative.[50]

Interestingly, exchange of the iodoacrylate to the bromo derivative in **4.38.** leads to a different reaction. Dimerization of two hexyne molecules and subsequent attachment of the acrylate gave 2-(2',3',4',5'-tetraethylcyclopentadienyl)-acetic acid ethyl ester in 74% yield.[51]

$$Et\!\!-\!\!\!\equiv\!\!\!-\!\!Et \; + \; \underset{\text{CO}_2\text{Me}}{\overset{\text{I}}{\diagup}} \quad \xrightarrow[\text{Zn, MeCN, 50\%}]{\text{Ni(PPh}_3)_2\text{Cl}_2} \quad \overset{\text{Et}}{\underset{\text{Et}}{\diagdown}}\text{pyrone} \qquad (4.38.)$$

In the coupling of Z-3-iodo-3-trimethylsilylacrylic acid and allenyltin reagents the carbon-carbon bond was established in a Stille coupling, followed by the attack of the carboxylic acid on the allene moiety, probably promoted by palladium (**4.39.**).[52]

$$\underset{\text{I}\quad\text{OH}}{\overset{\text{Me}_3\text{Si}}{\diagdown}}\text{O} \; + \; \text{Me}\diagdown\diagdown\text{SnBu}_3 \xrightarrow[\text{Na}_2\text{CO}_3,\,\text{DMF, 86\%}]{\text{Pd(OAc)}_2,\,\text{PPh}_3} \quad \overset{\text{SiMe}_3}{\underset{\text{Et}\quad\text{O}\quad\text{O}}{}} \qquad (4.39.)$$

The coupling of 2-iodobenzoic acid and phenylacetylene under Sonogashira coupling conditions was found to give a mixture of an isocoumarin derivative and a phthalide (**4.40.**). The proper choice of the catalyst system led to the preferential formation of the latter compound.[53] The process might also be diverted towards the formation of the isocoumarin derivative by isolation of the intermediate *o*-ethynyl-benzoic acid and its subjection to carefully selected cyclization conditions.[54]

$$\underset{\text{CO}_2\text{H}}{\overset{\text{I}}{\diagup}} \; + \; \underset{\text{Ph}}{|||} \xrightarrow[\text{Et}_3\text{N, MeCN, 36\%}]{\text{Pd(PPh}_3)_2\text{Cl}_2,\,\text{CuI}} \quad \text{isocoumarin} \; + \; \text{phthalide} \qquad (4.40.)$$

$$6 \; : \; 4$$

The analogous "open chain" carboxylic acid, Z-non-2-en-4-ynoic acid, when treated with 4-iodoanisole in the presence of a palladium-triphenylphosphine catalyst and potassium carbonate gave a mixture of three products, two of which were isolated (**4.41.**): *i*) the pyrone derivative arising from the attack of the anisylpalladium complex at the 4-position, followed by ring closure; *ii*) the furane derivative (major product) arising from the attack of the anisylpalladium complex at the 5-position, followed by ring closure and; *iii*) the pyrone derivative arising from the palladium catalyzed ring closure of the starting material (not always observed).[55]

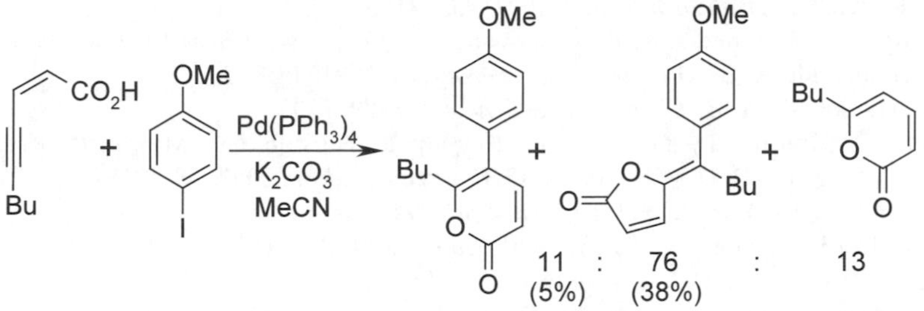

$$11 \quad : \quad 76 \quad : \quad 13$$
$$(5\%) \quad (38\%)$$

(4.41.)

Trost exploited the annulation of electron rich phenols and alkynoates to obtain coumarins in the presence of transition metal complexes. Ethyl propiolate and 3,4,5-trimethoxyphenol were coupled in formic acid in the presence of a palladium complex and sodium acetate to give 5,6,7-trimethoxycoumarin via a net C-H insertion in acceptable yield (**4.42.**). The coupling, characteristic of electron rich phenols, was also catalyzed by other transition metals, such as platinum or silver.[56]

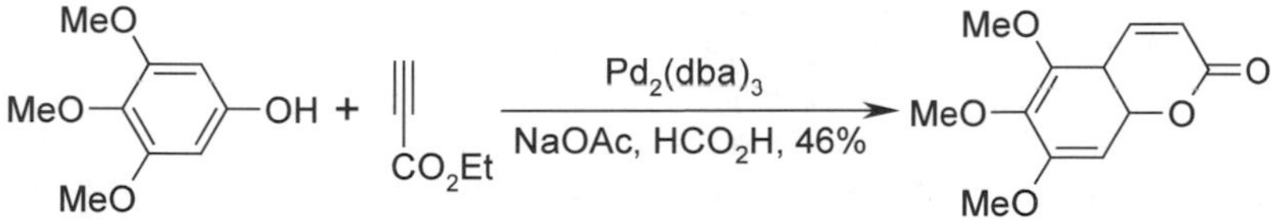

(4.42.)

The nickel catalyzed cyclization of 7-oxabenzonorbornadiene and methyl butyn-2-oate in the presence of a stoichiometric amount of zinc led to the formation of 4-methyl-benzocoumarin in good yield (**4.43.**). In the proposed mechanism of the process the reductive coupling of the starting materials gives a naphthylacrylic acid derivative, which under the applied conditions undergoes spontaneous *Z-E* isomerisation and ring closure to furnish the tricyclic product.[57]

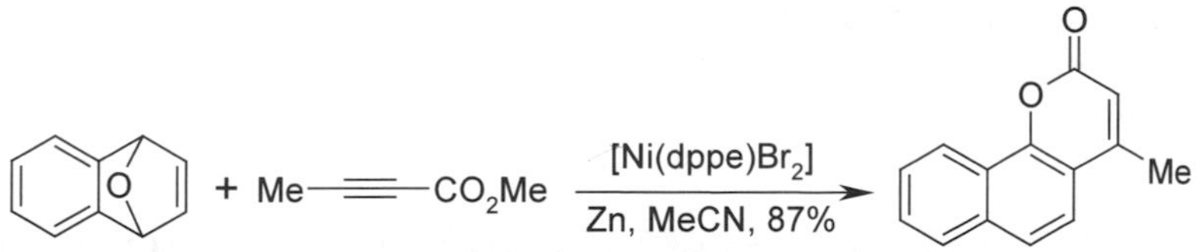

(4.43.)

4.5   REFERENCES

[1] Sakamoto, T.; Yasuhara, A.; Kondo, Y.; Yamanaka, H. *Heterocycles* **1993**, *36*, 2597.

[2] Hesse, S.; Kirsch, G. *Tetrahedron Letters* **2002**, *43*, 1213.

[3] Malm, J.; Rehn, B.; Hörenfeldt, A.-B.; Gronowitz, S. *J. Heterocycl. Chem.* **1994**, *31*, 11.

[4] Malm, J.; Hörenfeldt, A.-B.; Gronowitz, S. *Heterocycles* **1993**, *35*, 245.

[5] Zhang, Y.; Hörenfeldt, A.-B.; Gronowitz, S. *Synthesis* **1989**, 130.

[6] Tapolcsányi, P.; Krajsovszky, G.; Andó, R.; Lipcsey, P.; Horváth, Gy.; Mátyus, P.; Riedl, Zs.; Haós, Gy.; Maes, B. U. W.; Lemiére, G. L. F. *Tetrahedron*, **2002**, *58*, 10137.

[7] Solé, D.; Diaba, F.; Bonjoch, J. *J. Org. Chem.* **2003**, *68*, 5746.

[8] Solé, D.; Vallverdú, L.; Bonjoch, J. *Adv. Synth. Catal.* **2001**, *343*, 459.

[9] Gilchrist, T.L.; Kemmitt, P.D.; Germain, A.L. *Tetrahedron* **1997**, *53*, 4447.

[10] Sánchez-Sancho, F.; Mann, E.; Herradón, B. *Adv. Synth. Catal.* **2001**, *343*, 560.

[11] Kim, G.; Kim, J. H.; Kim, W.-j.; Kim, Y. A. *Tetrahedron Lett.* **2003**, *44*, 8207.

[12] Knölker, H.-J.; Filali, S. *Synlett* **2003**, 1752.

[13] Ma, S.; Ni, B. *J. Org. Chem.* **2002**, *67*, 8280.

[14] Zhang, H.; Larock, R. C. *J. Org. Chem.* **2003**, *68*, 5132.

[15] Kadnikov, D. V.; Larock, R. C. *J. Org. Chem.* **2004**, *69*, 6772.

[16] (a) Kalinin, V. N.; Shostakovsky, M. V.; Ponomaryov, A. B. *Tetrahedron Lett.* **1990**, *31*, 4073. (b) Ciattini, P. G.; Morera, E.; Ortar, G.; Rossi, S. S. *Tetrahedron* **1991**, *47*, 6449. (c) Torri, S.; Okumoto, H.; Xu, L.-H.; Sadakane, M.; Shostakovsky, M. V.; Ponomaryov, A. B.; Kalinin, V. N. *Tetrahedron* **1993**, *49*, 6773.

[17] Miao, H.; Yang, Z. *Org. Lett.* **2000**, *2*, 1765.

[18] Kadnikov, D. N.; Larock, R. C. *Org. Lett.* **2000**, *2*, 3643.

[19] Okuro, K.; Alper, H. *J. Org. Chem.* **1997**, *62*, 1566.

[20] Kwong, F. Y.; Buchwald, S. *Org. Lett.* **2003**, *5*, 793.

[21] Abouabdellah, A.; Dodd, R. H. *Tetrahedron Lett.* **1998**, *39*, 2119.

[22] Kuwabe, S.-i.; Torraca, K. E.; Buchwald, S. L. *J. Am. Chem. Soc.* **2001**, *123*, 12202.

[23] Trost, B. M.; Shen, H. C.; Dong, L.; Surivet, J.-P. *J. Am. Chem. Soc.* **2003**, *125*, 9276.

[24] Tanimori, S.; Kato, Y.; Kirihata, M. *Synthesis* **2004**, *13*, 2103.

[25] Malapel-Andrieu, B.; Mérour, J.-Y.; *Tetrahedron* **1998**, *54*, 11095.

[26] Watanabe, K.; Hiroi, K. *Heterocycles* **2003**, *59*, 453.

[27] Maezaki, N.; Yagi, S.; Maeda, J.; Yoshigami, R.; Tanaka, T. *Heterocycles* **2004**, *62*, 263.

[28] Kato, K.; Ono, M.; Akita, H. *Tetrahedron Lett.* **1997**, *38*, 1805.

[29] Costa, M.; Ca, N. D.; Gabriele, B.; Massera, C.; Salerno, G.; Soliani, M. *J. Org. Chem.* **2004**, *69*, 2469.

[30] Zhu, J.; Germain, A. R.; Porco, J. A. Jr. *Angew. Chem. Int. Ed.* **2004**, *43*, 1239.

[31] Patil, N. T.; Yamamoto, Y. *J. Org. Chem.* **2004**, *69*, 5139.

[32] Kuroda, T.; Suzuki, F. *Tetrahedron Lett.* **1991**, *32*, 6915.

[33] Miki, Y.; Shirokishi, H.; Matsushita, K. *Tetrahedron Lett.* **1999**, *40*, 4347.

[34] Harayama, T.; Toko, H.; Kubota, K.; Nishioka, K.; Abe, H.; Takeuchi, Y. *Heterocycles* **2002**, *58*, 175.

[35] Harayama, T.; Toko, H.; Nishioka, H.; Abe, H.; Takeuchi, Y. *Heterocycles*, **2003**, *59*, 541.

[36] Harayama, T.; Toko, H.; Hori, A.; Miyagoe, T.; Sato, T.; Nishioka, K.; Abe, H.; Takeuchi, Y. *Heterocycles* **2003**, *61*, 513.

[37] Harayama, T.; Sato, T.; Hori, A.; Abe, H.; Takeuchi, Y. *Synlett* **2003**, 1141.

[38] Melnyk, P.; Gasche, J.; Thal, C. *Tetrahedron Lett.* **1993**, *34*, 5449.

[39] Melnyk, P.; Legrand, B.; Gasche, J.; Durcot, P.; Thal, C. *Tetrahedron* **1995**, *51*, 1941.

[40] Malapel-Andrieu, B.; Mérour, J.-Y. *Tetrahedron* **1998**, *54*, 11079.

[41] Black, D. St C.; Keller, P. A.; Kumar, N. *Tetrahedron* **1993**, *49*, 151.

[42] Roesch, K. R.; Zhang, H.; Larock, R. C. *J. Org. Chem.* **2001**, *66*, 8042.

[43] Zhang, H.; Larock, R. C. *Org. Lett.* **2001**, *3*, 3083.

[44] Ferraccioli, R.; Carenzi, D.; Rombolá, O.; Catellani, M. *Org. Lett.* **2004**, *6*, 4759.

[45] Zhang, H.; Larock, R. C. *J. Org. Chem.* **2002**, *67*, 7048.

[46] The alternate reaction sequence, starting from *o*-iodobenzaldehye imine and carrying out two subsequent transition metal catalyzed transformations was also reported: Roesch, K. R.; Larock, R. C. *Org. Lett.* **1999**, *1*, 553.

[47] Huang, Q.; Hunter, J. A.; Larock, R. C. *Org. Lett.* **2001**, *3*, 2973.

[48] Campo, M. A.; Huang, Q.; Yao, T.; Tian, Q.; Larock, R. C. *J. Am. Chem. Soc.* **2003**, *125*,

[49] Huang, Q.; Campo, M. A. ; Yao, T.; Tian, Q.; Larock, R. C. *J. Org. Chem.* **2004**, *69*, 8251.

[50] Larock, R. C.; Han, X.; Doty, M. J. *Tetrahedron Lett.* **1998**, *39*, 5713.

[51] Kotora, M.; Ishikawa, M.; Tsai, F.-Y.; Takahashi, T. *Tetrahedron* **1999**, *55*, 4969.

[52] Rousset, S.; Abarbri, M.; Thibonnet, J.; Duchêne, A.; Parrain, J.-L. *Chem. Commun.* **2000**, 1987.

[53] Kundu, N. G.; Pal, M.; Nandi, B.; *J. Chem. Soc., Perkin Trans. 1*, **1998**, 561.

[54] Bellina, F.; Ciucci, D.; Vergamini, P.; Rossi, R. *Tetrahedron* **2000**, *56*, 2533.

[55] Rossi, R.; Bellina, F.; Biagetti, M.; Catanese, A.; Mannina, L. *Tetrahedron Lett.* **2000**, *41*, 5281.

[56] Trost, B. M.; Toste, F. D.; Greenman, K. *J. Am. Chem. Soc.* **2003**, 125, 4518.

[57] Rayabarapu, D. K.; Sambaiah, T.; Cheng, C.-H. *Angew. Chem. Int. Ed.* **2001**, *40*, 1286.

Chapter 5

THE SYNTHESIS OF OTHER RING SYSTEMS

The transition metal catalyzed synthesis of seven membered and larger heterocycles attracted considerably less attention than the preparation of their five and six membered analogues. Typical examples in this chapter include the formation of heterocycles in insertion reactions, or through carbon-heteroatom bond formation. Although the formation of some macrocyclic natural products was also achieved in cross-coupling reactions they will not be discussed in detail.

5.1   TRANSMETALATION ROUTE

The formation of seven membered heterocycles and larger rings through cross-coupling reactions is quite rare (except for some macrocyclic natural products). An example of such a process is presented in **5.1**. The intramolecular Stille-coupling of the tributylstannyl-indole and vinyl bromide moieties led to the formation of a seven membered ring in good yield.[1]

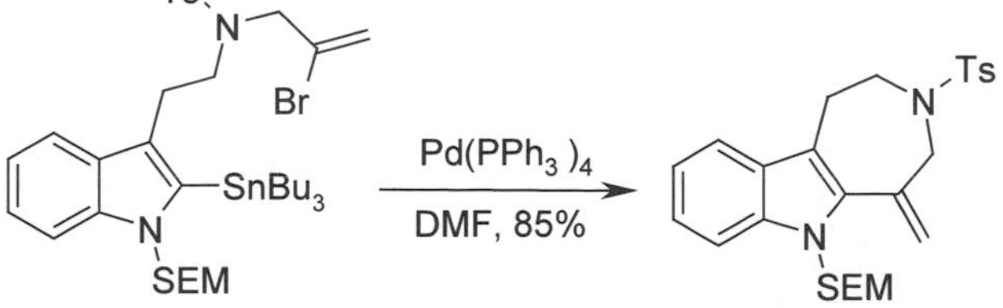

(5.1.)

Equation **5.2.** describes a transformation, where the palladium catalysed Suzuki coupling was used to establish the first carbon-carbon bond between an indole ring and an acetophenone moiety en route to benzo[*c*]carbazole. The carbacycle was established in a potassium *tert*-butoxide mediated photochemical transformation.[2]

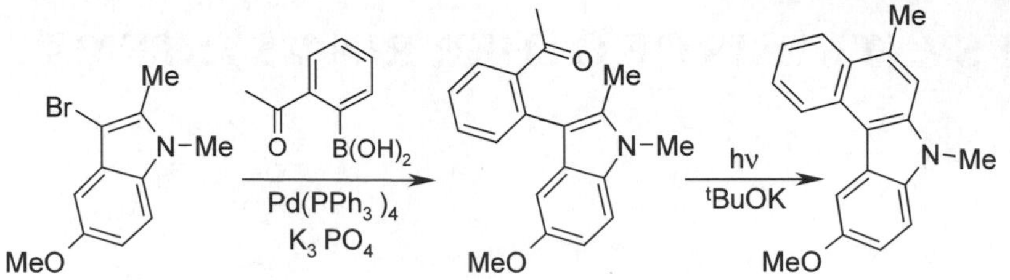

(5.2.)

5.2   INSERTION ROUTE

Transition metal catalyzed insertion reactions offer a variety of alternate approaches for the preparation of heterocyclic rings, of which Heck reactions were utilised extensively to prepare rings with more than 6 atoms. At the end of this Chapter some examples of the use of insertion reactions in the formation of the carbacyclic part of condensed heterocyclic systems will also be presented.

Heck reaction (olefin insertion)

The pentacyclic framework of natural product maxonine was prepared in an intramolecular 7-*exo* Heck cyclization. The migratory insertion of the pendant olefin into the arylpalladium complex could have led either to the formation of an eight or a seven membered ring, of which only the latter was observed (**5.3.**).[3]

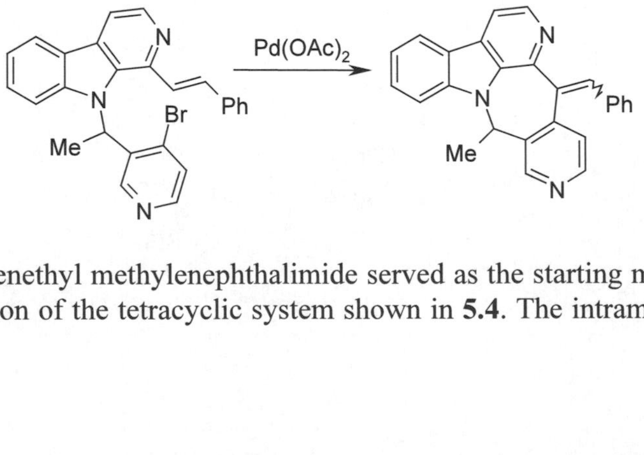

(5.3.)

N-phenethyl methylenephthalimide served as the starting material for the preparation of the tetracyclic system shown in **5.4**. The intramolecular Heck reaction led to selective carbon-carbon bond formation between the exocyclic carbon atom of the olefin bond and the aryl moiety.[4]

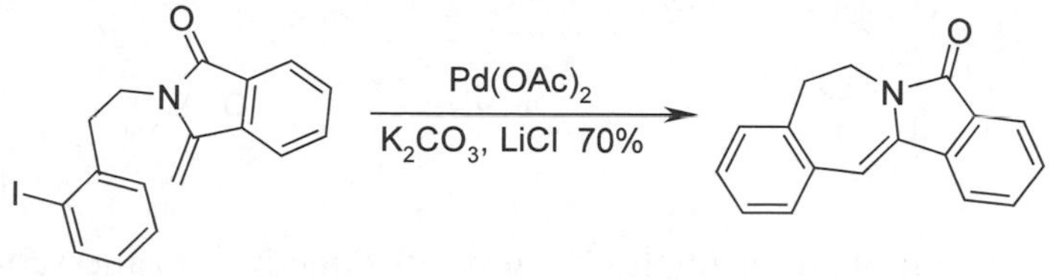

(5.4.)

The seven membered core of iboga alkaloids has also been constructed in an intramolecular Heck reaction. Insertion of a pendant olefin into the indolylpalladium complex, formed from iodoindole, followed by β-hydride elimination gave the complex framework of the natural product (**5.5.**). Although the insertion step could have led to the formation of a six membered ring, the formed palladium complex would have contained a quaternary carbon center in the β-position, blocking the closure of the catalytic cycle under the applied conditions.[5]

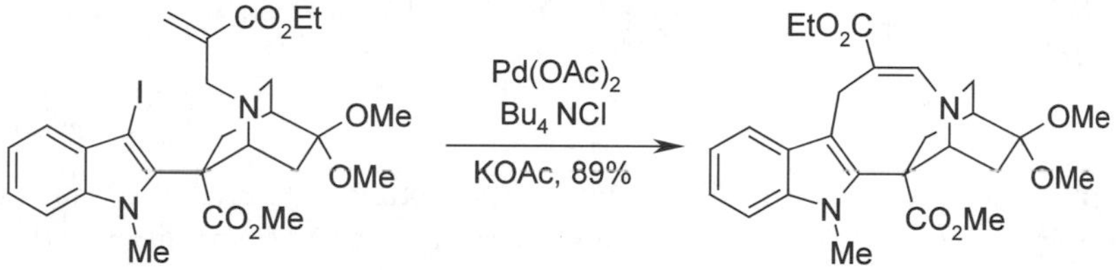

(5.5.)

The intramolecular Heck cyclization of resin bound *N*-homoallyl-2-iodobenzamides has been the key step in the preparation of substituted benzazepines (**5.6.**). The ring closure has also been extended to internal akynes, which gave benzylidene-benzazepines.[6]

In an analogous process polystyrene bound aryl halides, containing a pendant double bond were ring closed, giving the seven membered lacton as the major product (**5.7.**)[7]

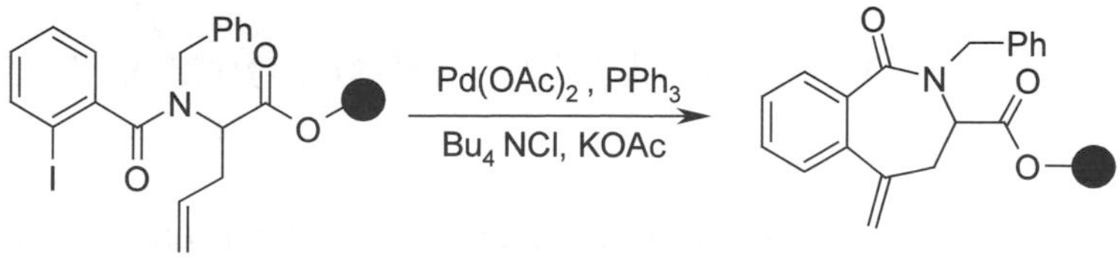

(5.6.)

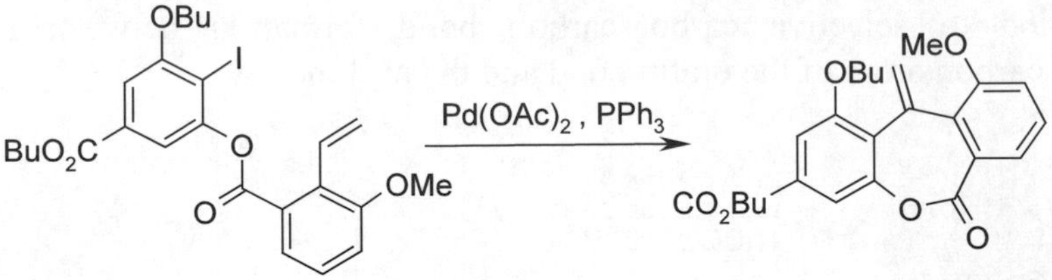

$$(5.7.)$$

Although the following examples do not fall directly into the scope of this chapter, they are discussed here since the key transformation in both of them is a transition metal catalysed insertion.

The first reaction is unique in a sense that both a five and a six membered heterocycle are formed in a tandem Heck reaction step. Closure of a five membered ring followed by the closure of the six membered ring converts the enamine shown in **5.8.** to the isoindoloneisoquinoline framework.[8]

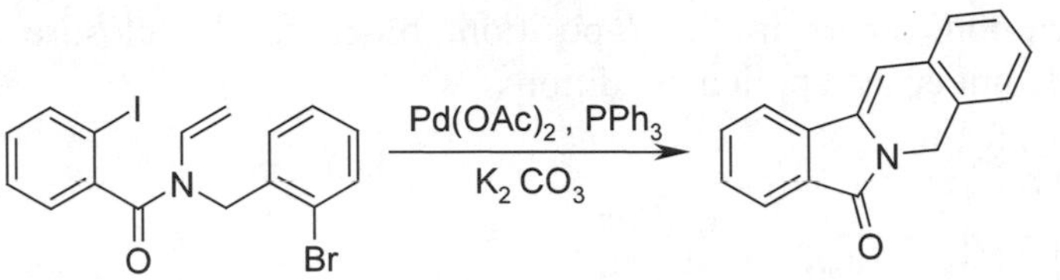

$$(5.8.)$$

In the second example carbon monoxide insertion into a benzothiophenylpalladium complex, and a subsequent C-H activation led to the transformation of 3-iodo-2-phenyl-benzothiophene to the tetracyclic compound shown in **5.9.**[9]

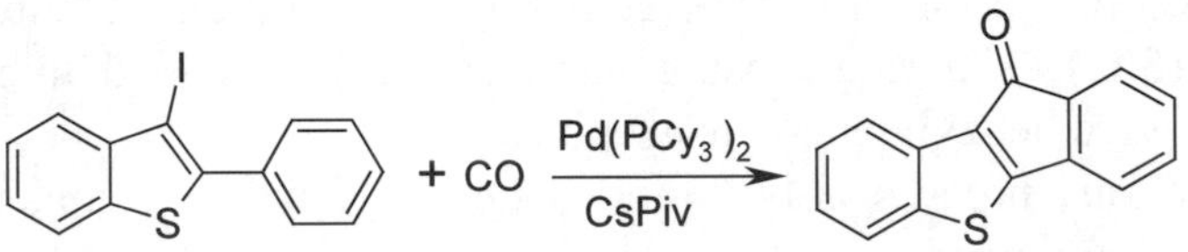

$$(5.9.)$$

5.3   CARBON-HETEROATOM BOND FORMATION

It is not surprising, that the transition metal catalyzed carbon-heteroatom bond forming reactions constitute a major part of the transformations used for the preparation of heterocyclic systems. This chapter summarises all those catalytic transformations, where the heterocycle is formed in the intramolecular connection of a carbon atom and a heteroatom.

The palladium catalyzed carbon-oxygen bond forming reaction was utilised by Buchwald and co-workers in the synthesis of a series of oxygen containing heterocycles including benzoxepins. Starting either from an *ω*-(*o*-halophenyl)-butanol or the analogous pentan-2-ol (**5.10.**), the formation of the seven membered ring was achieved in good yield.[10]

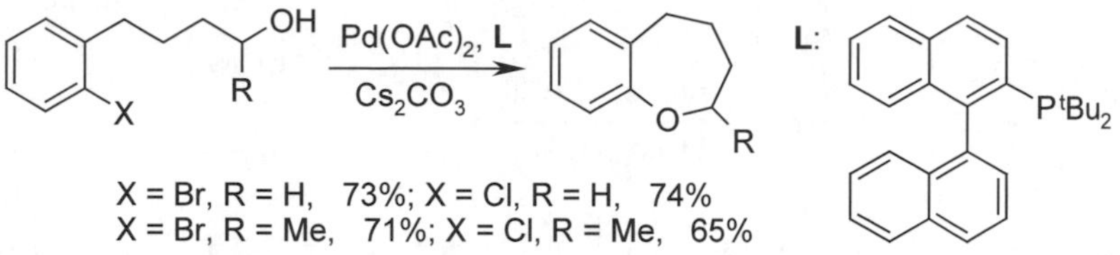

OH
R
X
Pd(OAc)2, L
Cs2CO3
O
R
L:
PtBu2
X = Br, R = H, 73%; X = Cl, R = H, 74%
X = Br, R = Me, 71%; X = Cl, R = Me, 65%

(5.10.)

The construction of the dibenzoxepino[4,5-*d*]pyrazole ring system was completed by the closure of the central oxepine ring. Attempts at the ring closure under both palladium catalysed and copper mediated conditions revealed, that each method has its advantages and limitations (**5.11.**). While the Buchwald-Hartwig coupling offers the use of only catalytic amounts of palladium, in contrast to the use of an equimolar amount of copper in the Ullman coupling, at the same time it is more sensitive to the steric bulk of the coupling partners.[11]

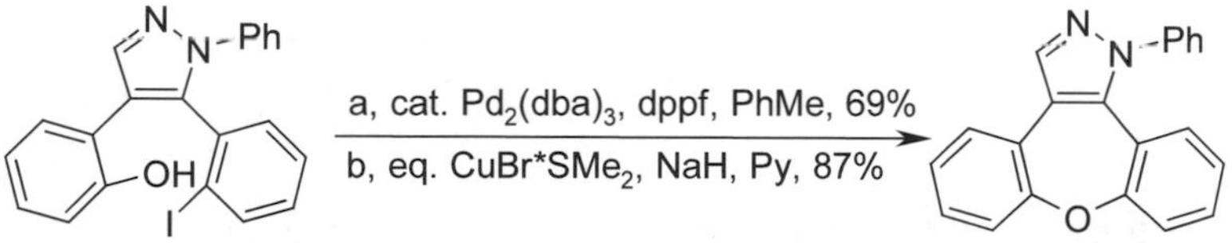

N
N-Ph
OH
I
a, cat. Pd2(dba)3, dppf, PhMe, 69%
b, eq. CuBr*SMe2, NaH, Py, 87%
O

(5.11.)

The transition metal catalysed nucleophilic attack of heteroatoms onto triple bonds has also been exploited in the preparation of larger rings. The intramolecular reaction of the phenol derivative shown in **5.12.** and the pendant triple bond could result in the formation of a seven or an eight membered ring. Although the former system would be favoured by geometric considerations, the only product formed in the process is the benzoxazocin derivative, whose formation is probably driven by electronic factors.[12]

OH
N
Ph
O
MeO
Pd(OAc)2 , LiCl
K2CO3, 62%
Ph
O
N
O
OMe

(5.12.)

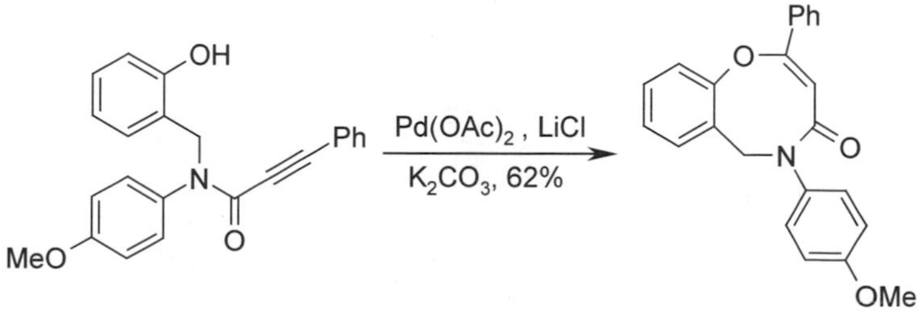

Sulfoxides bearing a pendant alkynyl group were found to undergo *syn*-selective sulfinylzincation. Palladium catalyzed diethylzinc mediated dealkynylation of the *tert*-butylethynyl-sulfoxides (**5.13.**), followed by a spontaneous ring closure led to the formation of a series of different sulphur containing heterocycles, including the seven membered ring.[13]

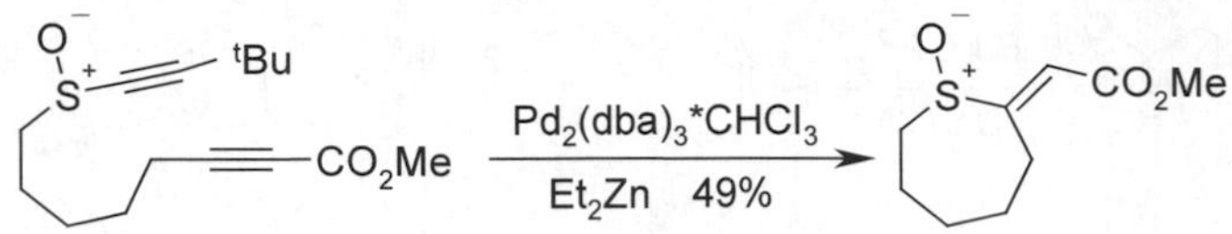

(5.13.)

The palladium catalyzed intramolecular nucleophilic substitution of allyl alcohol derivatives (Tsuji-Trost reaction) has successfully been extended to the closure of a seven membered ring. The coupling of the allyl alcohol unit and the enamide was the key step in the preparation of the natural product claviciptic acid (**5.14.**).[14]

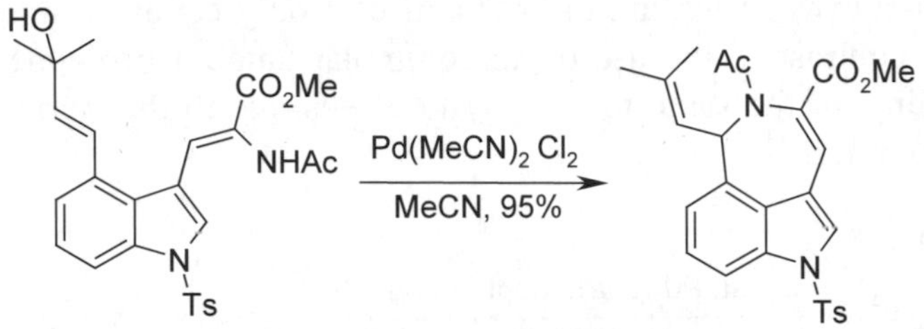

(5.14.)

The palladium catalysed addition of N-H or O-H bonds onto allenes has successfully been exploited in the preparation of oxazepines, diazepines, oxazocines and diazocines. The nucleophilic attack of the pendant alcohol or sulphonamide function on the allene moiety was followed by the incorporation of the alcohol, used as solvent, to give the desired cyclic products in good yield (**5.15.** and **5.16.**). The bromoallene in these processes is the synthetic equivalent of an allylic dication.[15]

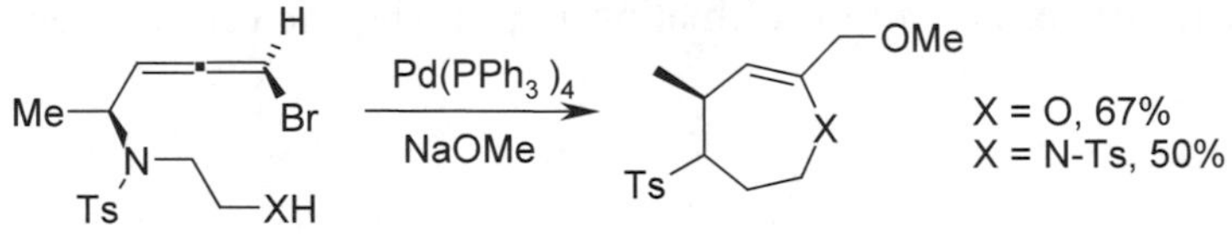

X = O, 67%
X = N-Ts, 50%

(5.15.)

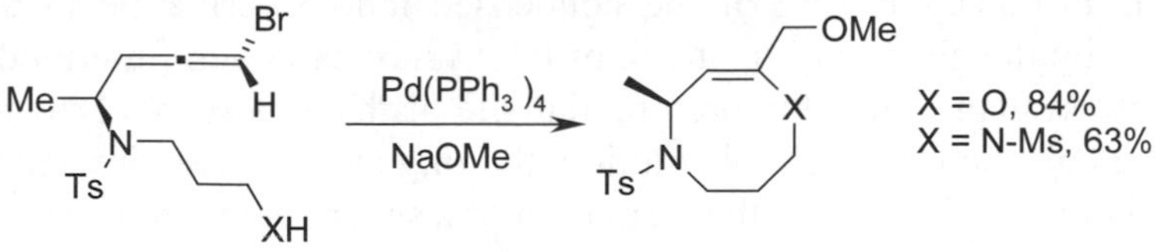

(5.16.)

5.4   OTHER PROCESSES

The examples described in this chapter achieve the formation of a seven membered ring system through C-H activation or oxidative coupling.

Tethering an iodobenzene and a furane moiety through a β-lactam led to the formation of a unique tetracyclic β-lactam derivative (**5.17.**). Oxidative addition followed by the carbopalladation of the furane ring resulted in the closure of the seven membered core in a so called "heteroaryl Heck reaction".[16]

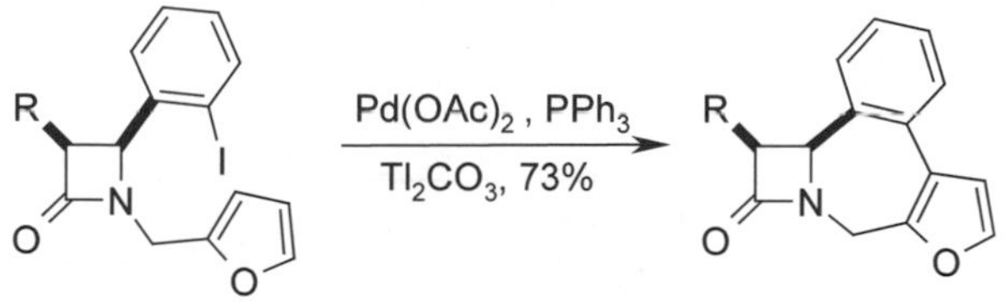

(5.17.)

The palladium catalysed ring closure of the *N*-(2'-bromobenzyl)-naphthylamine derivative shown in **5.18.** could in principle occur either in the 2-position or in the 8-position of the naphthalene ring giving rise to the formation of a six or a seven membered ring respectively. Surprisingly the latter case is favoured in most cases, unless there is a bulky substituent in position 7 of the naphthalene ring. This finding, where preference of the observed reaction path is explained by the intramolecular coordination of the benzylamino function to the palladium, has been successfully exploited in the preparation of a series of naphthobenzazepines.[17]

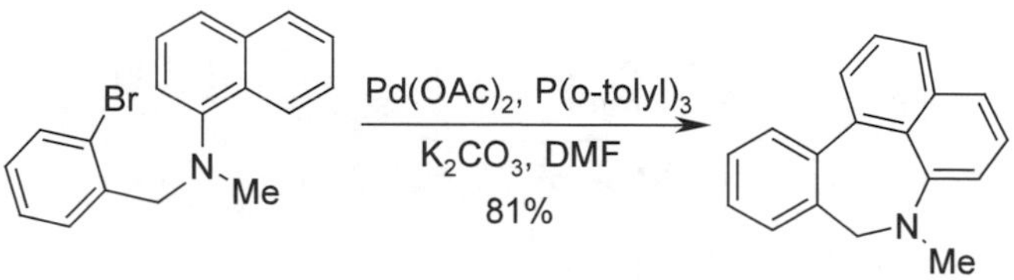

(5.18.)

The central diazepine ring of the condensed indole derivative in **5.19.** was built up by the electrophilic attack of the pyridylpalladium intermediate on the indole system. Since the preferential site of the attack, the 2-position of indole was blocked through substitution, the ring closure was directed into the *peri*-position, giving rise to the formation of a seven membered ring.[18]

$$(5.19.)$$

The palladium catalysed sequential alkylation-alkenylation of 5-iodoquinoline leads to the formation of the quinolooxepin ring system (**5.20.**). The process, closely related to the Catellani reaction,[19] runs through an *ortho*-alkylation – Heck reaction sequence. The preparation of a series of benzoxepines has also been achieved in this manner, starting from such iodobenzene derivatives, where one of the *ortho*-positions was blocked by substitution.[20]

$$(5.20.)$$

Besides the synthetic route depicted in **5.5.** the seven membered central heterocyclic ring of iboga alkaloids was also accessed by the palladium mediated closure of the indole derivative bearing a pendant double bond shown in **5.21**. The reaction is believed to be initiated by the palladation of indole in the 2-position followed by insertion of the double bond. Since β-hydride elimination from the formed intermediate is blocked by the conformational rigidity of the system, the reaction stops at this stage and the addition of a reducing agent is required to remove the palladium and obtain the product.[21]

$$(5.21.)$$

In spite of the facts that the heterocycles formed in the reactions depicted in **5.22.** are only of medium size, and the reactions are stoichiometric in palladium they deserve mentioning since a formally minor change in the reagent, exchange of an oxygen to nitrogen, results in a marked alteration of the reaction path. Both reactions are believed to commence by the *endo*-carbopalladation of norbornadiene by the arylpalladium complex. In the case of phenol attachment of the oxygen to the palladium and reductive elimination lead to the benzofurane derivative. In case of the aniline derivative the palladium complex is rearranged before the attachment of the nitrogen to the palladium takes place, opening up the way to the reductive elimination and closure of the six membered ring.[22]

(5.22.)

5.5   REFERENCES

[1] Palmsano, G.; Santagostino, M. *Synlett* **1993**, 771.

[2] de Koning, C. B.; Michael, J. P.; Nhlapo, J. M.; Pathak, R.; van Otterlo, W. A. L. *Synlett* **2003**, *5*, 705.

[3] Kelly, T. R.; Xu, W.; Sundaresan, J. *Tetrahedron Lett.* **1993**, *34*, 6137.

[4] Kim, G.; Kim, J. H.; Kim, W.-j.; Kim, Y. A. *Tetrahedron Lett.* **2003**, *44*, 8207.

[5] Sundberg, R. J.; Cherney, R. J. *J. Org. Chem.* **1990**, *55*, 6028.

[6] Bolton, G. L.; Hodges, J. C. *J. Comb. Chem.* **1999**, *1*, 130.

[7] Denieul, M.-P.; Skydstrup, T. *Tetrahedron Lett.* **1999**, *40*, 4901.

[8] García, A.; Rodríguez, D.; Castedo, L.; Saá, C.; Domínguez, D. *Tetrahedron Letters* **2001**, *42*, 1903.

[9] Campo, M. A.; Larock, R. C. *J. Org. Chem.* **2002**, *67*, 5616.

[10] Kuwabe, S.-i.; Torraca, K. E.; Buchwald, S. L. *J. Am. Chem. Soc.* **2001**, *123*, 12202.

[11] Olivera, R.; SanMartin, R.; Churruca, F.; Domínguez, E. *J. Org. Chem.* **2002**, *67*, 7215.

[12] Davion, Y.; Guillaumet, G.; Léger, J.-M.; Jarry, C.; Lesur, B.; Mérour, J.-Y. *Heterocycles* **2003**, *60*, 1793.

[13] Maezaki, N.; Yagi, S.; Maeda, J.; Yoshigami, R.; Tanaka, T. *Heterocycles* **2004**, *62*, 263.

[14] Harrington, P. J.; Hegedus, L. S.; McDaniel, K. F. *J. Am. Chem. Soc.* **1987**, *109*, 4335.

[15] Ohno, H.; Hamaguchi, H.; Ohata, M.; Tanaka, T. *Angew. Chem.* **2003**, *115*, 1791.

[16] Burwood, M.; Davies, B.; Diaz, I.; Grigg, R.; Molina, P.; Sridharan, V.; Hughes, M. *Tetrahedron Lett.* **1995**, *36*, 9053.

[17] Harayama, T.; Sato, T.; Hori, A.; Abe, H.; Takeuchi, Y. *Synlett* **2003**, 1141.

[18] Melnyk, P.; Legrand, B.; Gasche, J.; Durcot, P.; Thal, C. *Tetrahedron* **1995**, *51*, 1941.

[19] Catellani, M.; Fagnola, M. C. *Angew. Chem. Int. Ed.* **1994**, *33*, 2421.

[20] Lautens, M.; Paquin, J.-F.; Piguel, S. *J. Org. Chem.* **2002**, *67*, 3972.

[21] Trost, B. M.; Godleski, S. A.; Genet, J. P. *J. Am. Chem. Soc.* **1978**, *100*, 3930.

[22] Saito, K.; Ono, K.; Sano, M.; Kiso, S.; Takeda, T. *Heterocycles* **2002**, *57*, 1781.

Chapter 6

THE FUNCTIONALIZATION OF FIVE MEMBERED RINGS

The transition metal catalyzed functionalization of five membered heterocycles constitutes one of the bigger chapters of this book. The large number of examples might be attributed to the following reasons: *i*) the ease of the preparation of halogenated azoles; *ii*) the convenient preparation of azolylmetal reagents through deprotonation or metal-halogen exchange. This second point is of particular importance since azolyl halides are considerably electron-rich, therefore to achieve efficient oxidative addition one usually has to use more active catalysts.

The biological activity of azoles and their derivatives (indoles, purines, etc.) and their abundance as structural motif in natural products made them a prime target and test ground in the development of catalytic transformations. This chapter is mainly dedicated to the reactions of monocyclic five membered heterocycles and indole. The chemistry of other condensed systems of importance, such as purines, is discussed in Chapter 8.

6.1   TRANSMETALATION ROUTE

One of the most frequently studied transition metal catalyzed transformations of azoles and indole is their participation in cross-coupling reactions. Due to the abundance of examples in this field we only present some representative examples of the different reaction classes. In this chapter reactions where a halogenated azole is used to introduce the five membered ring onto the palladium in the oxidative addition and processes, where the organometallic derivative of an azole is transferring its heterocycle onto the catalyst in a transmetalation step will be discussed parallel. The featured reactions were divided into subclasses along the commonly used name reactions.

Suzuki coupling

The availability of organoboronic acids and their functional group tolerance led to the emergence of Suzuki coupling as one of the most utilized cross-coupling reactions in recent years.[1] Usually both the coupling of haloazoles and organoboron compounds, or azolylboronic acids and aryl halides proceed well.[2]

2-Pyrroleboronic acid, prepared through the lithiation and borylation of N-protected pirroles, was shown to undergo Suzuki coupling with a series of aryl and heteroaryl halides (**6.1.**). The reactions are sometimes biased by the formation of bipyrrole as side product.[3] The same 2-arylpyrroles can also be prepared starting from the protected 2-bromopyrrole derivative and arylboronic acids as shown in **6.2.** These reactions usually give the desired coupling product in excellent yield.[4]

(6.1.)

(6.2.)

The bromo derivative of N-methylsuccinimide did also undergo Suzuki coupling when treated with naphthylboronic acid in the presence of palladium acetate, triphenylphosphine and potassium carbonate (**6.3.**). The coupled product was deprotected under the reaction conditions and an indole derivative was isolated in good yield, which was successfully converted into the hexacyclic naphthopyrrolo[3,4-c]carbazole structure. Using the analogous trimethylstannyl-naphthalene derivative and optimised Stille coupling conditions the desired product was isolated only in 56% yield.[5]

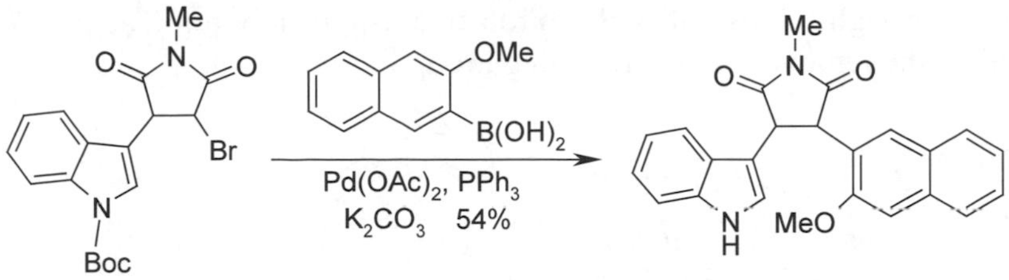

(6.3.)

The coupling of thiophene derivatives, particularly thiopheneboronic acids, is frequently employed in medicinal chemistry and also in materials science. In a recent example 2-thiopheneboronic acid was coupled with a triazolopyridine derivative (6.4.) in the presence of [1,1'-bis(diphenylphosphino)ferrocenyl]palladium dichloride in 83% yield.[6]

The coupling of the same boronic acid was also achieved with 4-chlorobenzoyl chloride (6.5.). Running the reaction under anhydrous conditions the desired 2-(4'chlorobenzoyl)thiophene was obtained in good yield.[7] The opening step in this process is the selective oxidative addition of the palladium into the carbonyl-chlorine bond giving an acylpalladium complex, which on subsequent transmetalation and reductive elimination gives the observed product.

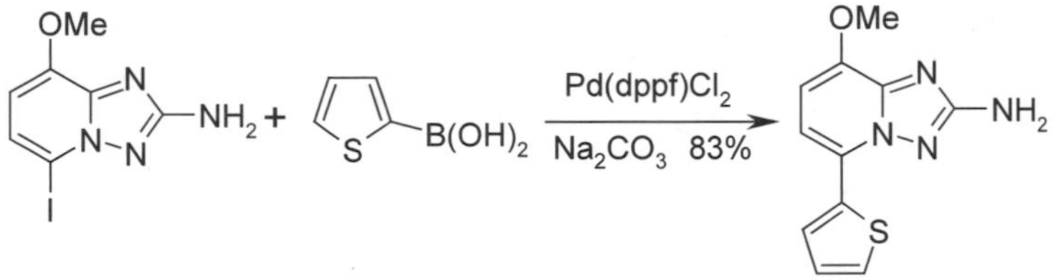

(6.4.)

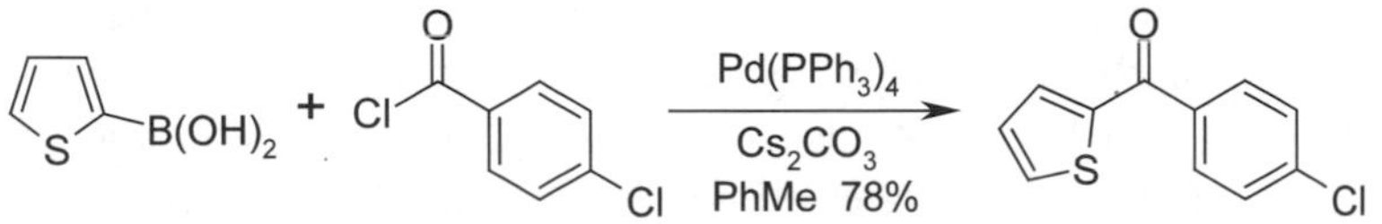

(6.5.)

The Suzuki coupling of thiopheneboronic acids is also frequently utilised in the preparation of oligomeric compounds. Sexithiophene (6.6.), a fluorescent material, was obtained by the coupling of the dibromo-quaterthiophene with 2-thiopheneboronic acid. Interestingly, the coupling failed under regular conditions, while microwave irradiation promoted the process efficiently to give the desired seximer in 65% yield.[8] The microwave assisted coupling was further extended and a solvent-free variant was also reported by the same authors recently.[9]

The key step in the preparation of a novel class of condensed thiophene based GABA receptor inverse agonists was the cross-coupling of the bromo derivative of the parent system. The 2-pyrrolyl substituent, for example, was introduced through the use of *N*-Boc protected 2-pyrroloboronic acid (**6.7.**) followed by the removal of the protecting group.[10]

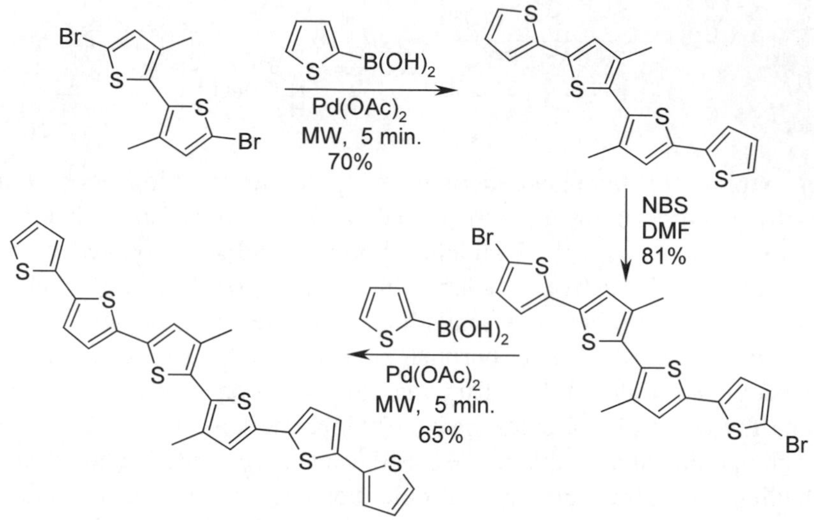

(6.6.)

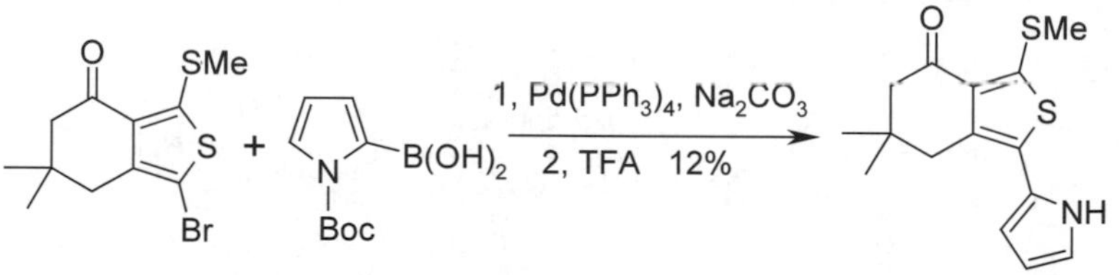

(6.7.)

The Suzuki coupling of thiopheneboronic acid also works with aryl triflates. In this case, however, the coupling gave better results using organic bases, than with carbonates. This behaviour was attributed to the increased sensitivity of the trifluoromethylsulfonyl group. Using triethylamine the indolyl-triflate (**6.8.**) was converted to the 3-thienyl derivative in good yield.[11]

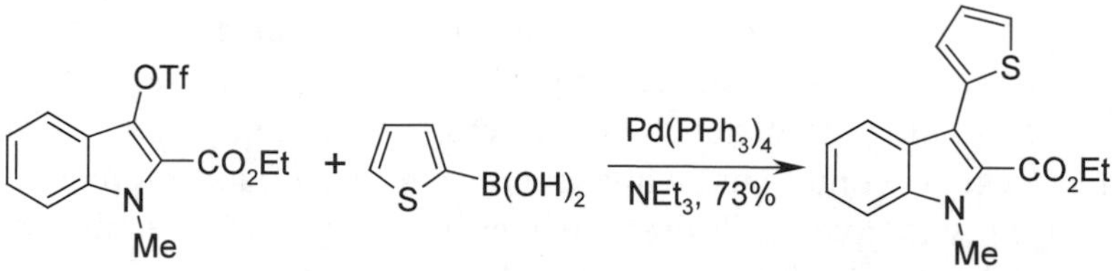

(6.8.)

A representative example of the coupling of furan derivatives is the preparation of the PDE5 inhibitor shown in **6.9.** The concluding step in the synthesis was the Suzuki coupling of the 2-bromofurane moiety with a series of arylboronic acids including 2-furylboronic acid. The couplings usually ran smoothly and the desired products were isolated in good to mediocre yield.[12]

In an analogous case 5-methyl-3-phenylisoxazole-4-boronic acid – prepared from the bromoisoxazole derivative in a lithium-bromine exchange, borylation sequence – was efficiently coupled with a series of aryl halides (**6.10.**). The coupling was run under conventional conditions employing tetrakis(triphenylphosphino)palladium for catalyst and sodium carbonate as base. Using this approach the preparation of valdecoxib, a highly potent COX-2 selective inhibitor, and its analogues was realised in good yield.[13]

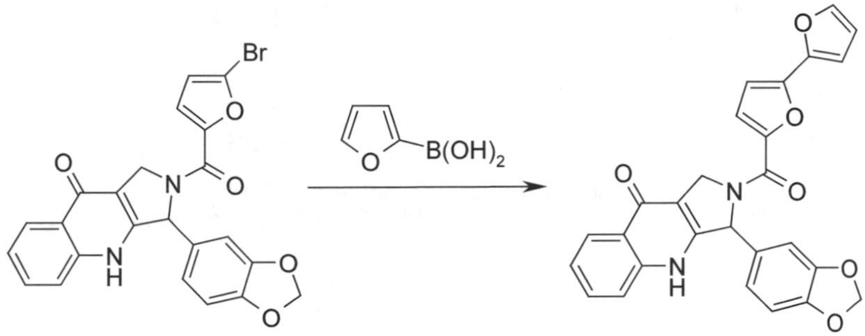

(6.9.)

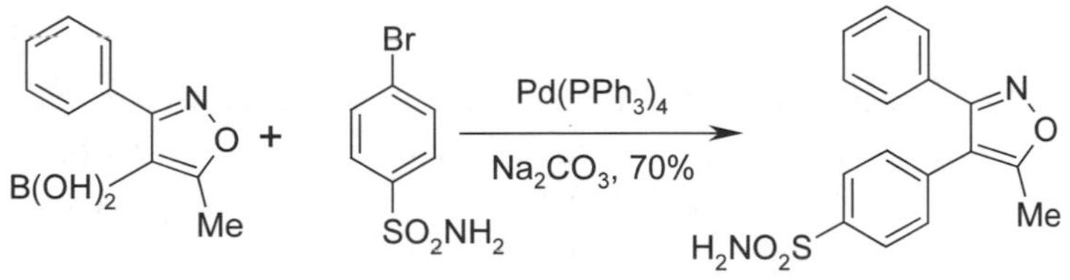

(6.10.)

Since the first report of an indoleboronic acid[14] by Gribble the Suzuki coupling has been widely employed to functionalise this ring system. Under Suzuki conditions the introduction of an aryl group onto the five membered ring can be achieved without *N*-protection as exemplified by the reaction of the 2-bromoindole derivative in **6.11.** with pyridineboronic acid[15] (*N.B.* most of the analogous coupling reactions require the use of *N*-protection).

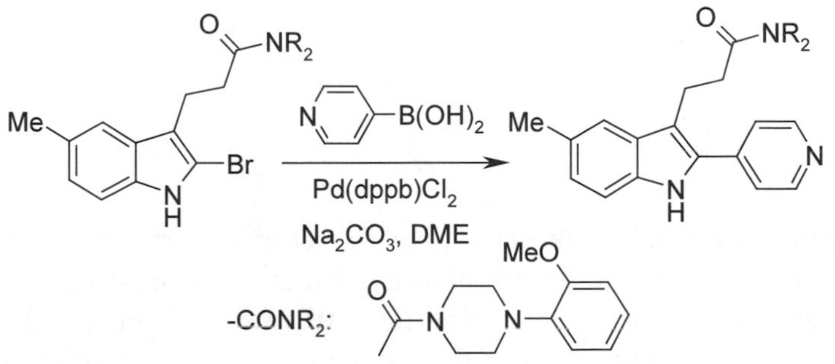

(6.11.)

The cross-coupling of 3,6-dibromoindole with arylboronic acids (**6.12.**) is a clear indication of the increased electron density of the five membered ring, since only the selective formation of 6-aryl-3-bromoindoles was observed. The selective introduction of the aryl group into the 3-position required a two-step approach. Consecutive conversion of this bromine into a boronic acid followed by the Suzuki coupling with aryl halides gave the desired 3-aryl-6-bromoindoles in a high overall yield. The success of the process relies heavily on the high selectivity of the lithium-bromine exchange and the Suzuki coupling.[16]

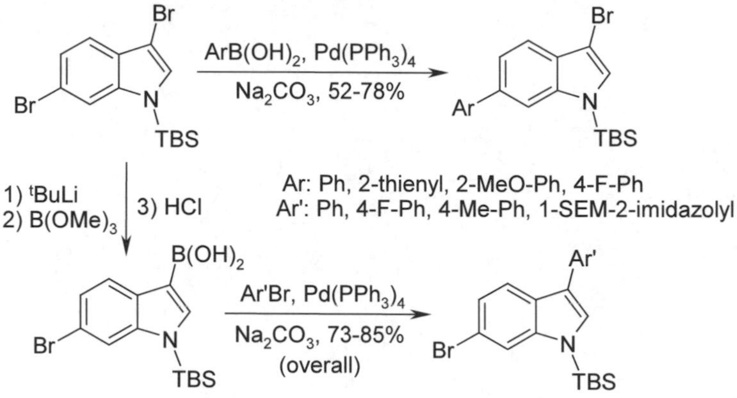

(6.12.)

The Suzuki coupling of aryl halides was also extended to tosylates recently. Benzothiazole 5-tosylate reacted with *m*-xylene-2-boronic acid (**6.13.**) to give the coupled product in 94% yield using palladium acetate and a stericly congested biphenyl based phosphine ligand as catalyst.[17] Another class of less commonly utilised cross-coupling partners are methyltio derivatives. In the presence of a copper salt, which activates the carbon-sulphur bond, 2-methyltio-benzotiazol coupled readily with a series of arylboronic acids.[18]

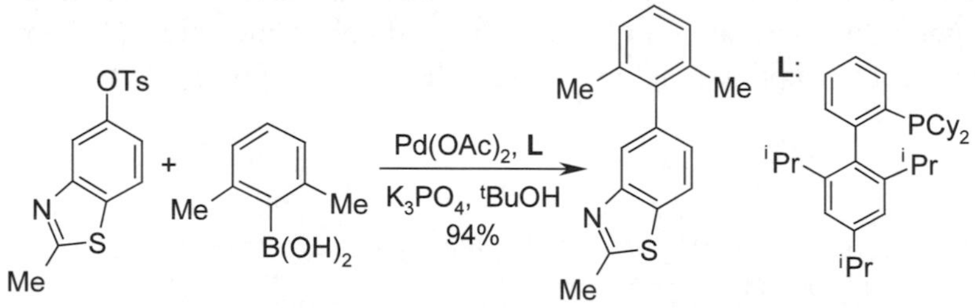

(6.13.)

The bis(indolo)pyrazine moiety is a structural motif that attracted attention due to its potent biological activity. An elegant construction of the frame reported by Jiang and co-workers utilizes the sequential cross-coupling of indole units and the pyrazine core. 2,5-Dibromo-3-

methoxypyrazine (**6.14.**) was coupled with an indole-3-boronic acid to introduce the indole moiety into the 3-position selectively. Any attempts to achieve the second cross-coupling of the so formed intermediate with an appropriately substituted indole-3-boronic acid derivative remained vain. The solution to this problem was finally provided by the use of the appropriate 3-trimethylstannyl-indole derivative, which gave the desired compound in92% yield.[19]

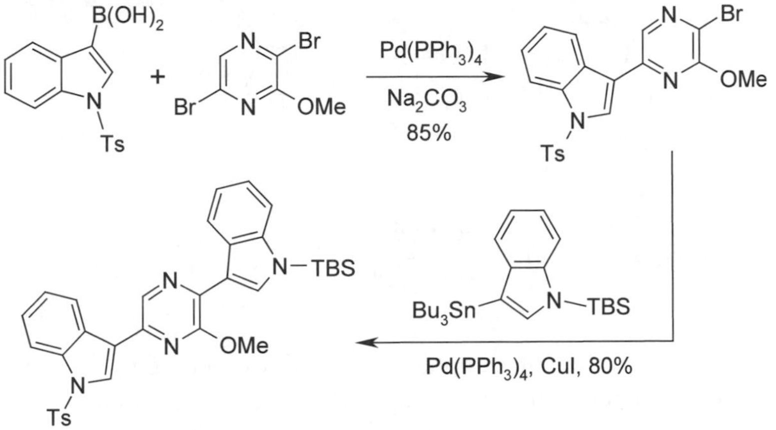

(6.14.)

Kharasch (Kumada) coupling

The palladium or nickel catalyzed cross-coupling of Grignard reagents and aryl or vinyl halides is commonly utilized in the functionalization of five membered heterocycles. The azoles might play the role of both the organometallic reagent and the halide, but examples of the former class will be presented in more detail, since the preparation of azolylmagnesium halides can conveniently be achieved both *via* the insertion of magnesium into an azole-halogen bond (C-X activation) or through the deprotonation of the azole followed by transmetalation with a magnesium halide (C-H activation).

Since the N-H bond of pyrrole is deprotonated by the strongly basic Grignard reagents, the Kumada coupling requires the *N*-protection of the pyrrole prior to the coupling. The triisopropylsilyl-protected 3-pyrrolomagnesium bromide (**6.15.**), prepared from the appropriate 3-bromopyrrole derivative and magnesium, was coupled in the presence of a palladium-dppf complex. Under these conditions different aryl and alkyl halides were found to couple efficiently to give the 3-aryl or 3-alkylpyrroles in good to excellent yield.[20]

Another demonstration of the efficiency of Kumada coupling in pyrrole chemistry is presented in **6.16.** 2-(*N*-methylpyrrolo)magnesium bromide and iodobenzene were coupled in excellent yield (93%) giving the desired *N*-methyl-2-phenypyrrole.[21] The palladium based catalyst contained (again) a bidentate phosphine ligand, in this case dppb. The Gringnard reagent in this case was prepared by the lithiation of *N*-methylpyrrole followed by a lithium-magnesium exchange.

(6.15.)

(6.16.)

The Kumada coupling of thiophene derivatives, providing convenient access amongst others to oligothiophenes and polythiophenes. was extensively studied, since most of these compounds have interesting electronic and optical properties.[22] The coupling of the bithienylmagnesium bromide in **6.17.** and 4-bromopyridine was run in the presence of a palladium-dppf catalyst in boiling diethyl ether to give 5-pyridyl-2,2'-bithienyl.[23]

(6.17.)

There are certain cases, however, when the Kumada coupling fails to give the desired product. Feringa and co-workers tried to couple bis(chlorothienyl)cyclopentene with 2-thienylmagnesium bromide at both termini to get an extended optical molecular switch.[24] The reaction, using a nickel-dppp catalyst, gave only the monocoupled product, shown in **6.18.**, in mediocre yield along some starting material. Increasing the amount of the catalyst to stoichiometric level resulted in no improvement and the change of the ligand to triphenylphosphine had only a minor effect on the reaction. The authors had to switch to the Suzuki-coupling to achieve the desired transformation.

Interestingly, the analogous double Kumada coupling of 2,3-dibromothiophene and 2 equivalents of a thienylmagnesium bromide derivative (**6.19.**) ran smoothly in the presence of nickel-dppp catalyst (*N.B.* the same catalyst failed to initiate double coupling in **6.18.!**).[25] The trithiophene derivative was isolated in 90% yield and used in subsequent cross-coupling steps to build thiophene dendrimers containing up to 30 heterocyclic rings.

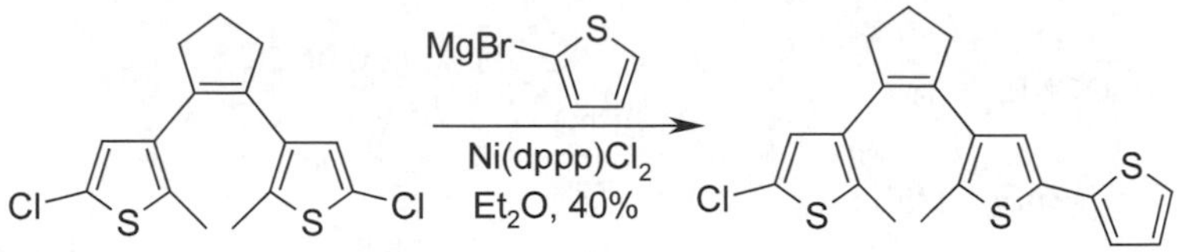

$$(6.18.)$$

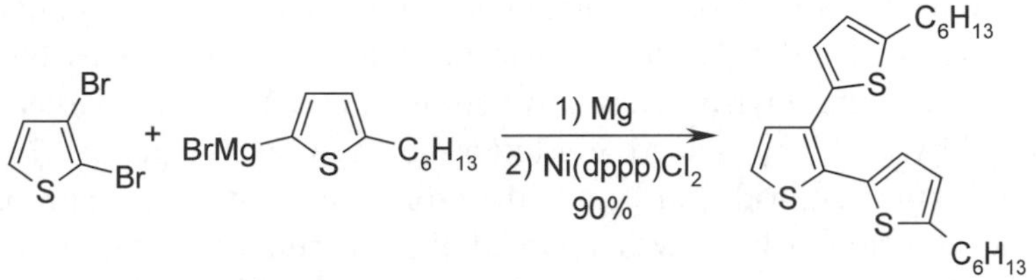

$$(6.19.)$$

The reaction sequence, shown in **6.20.**, was used to introduce two azolyl moieties onto the pyridine ring. In the first step, 2,5-dibromopyridine was reacted with an indolylmagnesium reagent in the presence of a palladium-dppb catalyst at ambient temperature to couple selectively at the more reactive 2-position and give the pyridylindole compound in 87% yield. In the second step, where the less reactive 5-position of the pyridine was arylated with 2-thienylmagnesium bromide under similar coupling conditions, the reaction mixture had to be heated to boiling to achieve the coupling and isolate the desired compound in an excellent 98% yield.

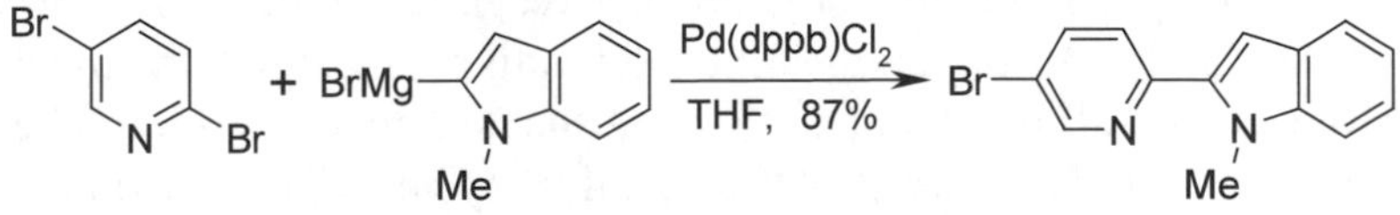

$$(6.20.)$$

Negishi coupling

The use of the mildly functional group tolerant but readily transmetalating organozinc derivatives of azoles has also found widespread application in heterocyclic chemistry. Their preparation is usually achieved by the transmetalation of the appropriate organolithium or Grignard reagent with an anhydrous zinc salt. This approach, coupled with directed *ortho*-lithiation or lithium-halogen exchange provides a convenient entry to functionalised biaryls.

Snieckus and co-workers reported the directed lithiation of 3-furanecarboxylic acid diethylamide (**6.21.**), which proceeded selectively in the 2-position, and the subsequent zinc-lithium exchange. The so formed furanylzinc reagent underwent Negishi-coupling with 2-bromotoluene in the presence of a palladium-triphenylphosphine catalyst to give 2-(*o*-tolyl)furane in good yield.[26]

(6.21.)

The scope of the Negishi-coupling is not limited to aryl and vinyl halides and sometimes acyl chlorides might also be converted to ketones by this protocol. The 2,3-dihalopyrrole derivative shown in **6.22.** was converted into its 2-lithio derivative by selective lithium-halogen exchange at -78 °C. Addition of zinc chloride effected the formation of the appropriate pyrrolylzinc chloride, which was coupled with a functionalised butyroyl chloride in the presence of tetrakis(triphenylphosphino)palladium and furnished the expected 2-acylpyrrole in 61% yield.[27]

(6.22.)

The lithium-halogen exchange was exploited in the synthesis and cross-coupling of thiophene-based organozinc reagents too. 4-Bromo-2-octylthiophene and its organometallic derivative 2-octyl-4-thienylzinc chloride were successfully coupled to give the symmetrical bithiophene (**6.23.**) using Fu's highly active Pd-P^{t}Bu$_3$ catalyst system. The reaction was also extended, although with limited success, to the introduction of two organozinc reagents onto the thiophene ring using 3,4-dibromothiophene as reagent.[28]

(6.23.)

The formation of 2,4'-bithiazoles, a structural motif found in a series of natural products exhibiting a diverse spectrum of biological activities, can also be achieved through the use of the Negishi coupling protocol. 4-Bromo-

2-isopropylthiazole was converted into its zinc derivative by Bach, which was successfully and selectively coupled with 2,4-dibromothiazole in the 2-position to give the product shown in **6.24**.[29] The preferred catalyst was formed *in situ* from $Pd_2(dba)_3$ and dppf, although in certain cases $Pd(PPh_3)_2Cl_2$ was also effective. It is interesting to note, that haloazoles also undergo magnesium-halogen exchange, which coupled with transmetalation to zinc and Negishi coupling was utilised in the preparation of pyrazole derivatives.[30]

(6.24.)

The selective deprotonation of *N*-protected indoles in the 2-position was utilised in the preparation of 2-uracylindole derivatives (**6.25.**). Reaction of *N*-methylindole with butyllithium followed by transmetalation with zinc chloride gave the 2-indolylzinc reagent, which was coupled with the silyl protected 5-iodouracyl to yield the desired product.[31] The preparation and coupling of the analogous 3-indolylzinc reagents is less straightforward, since approaches utilising the 3-indolyllithium intermediate are biased by its rearrangement to the thermodynamically more favoured 2-indolyllithium isomer. The observation, that TBDMS-protected 3-indolyllithium reagents are stable even at ambient temperature helped Bosch and co-workers to prepare a series of 3-arylindole derivatives starting from *N-tert*-butyldimtethylsilyl-3-bromoindole.[32]

(6.25.)

N-protected imidazole derivatives are usually deprotonated by strong bases selectively in the 2-position. This reaction, followed by transmetalation with zinc chloride, serves as the basis of the 2-arylation of the imidazole ring in the process reported by Bell and Ruddock.[33] The 2-

imidazolylzinc chloride derivative shown in **6.26.** was coupled with 2-bromopyridine to give 2-(2'-pyridyl)imidazole in 90% yield after loss of the protecting group during the acidic workup. The protocol also works well with similar systems, such as thiazole,[34] and benzthiazole.[35]

$$(6.26.)$$

Stille coupling

Prior to the recent emergence of the Suzuki and Hiyama reaction as functional group tolerant cross-coupling reactions, the Stille coupling was predominantly used for sensitive substrates. The relative stability of hetarylstannanes enables their isolation and use as a "shelf reagent". The common problems associated with Stille coupling, however, such as the toxicity of tri- and tetraorganotin reagents and the occasional difficulty encountered during workup, led in recent years to a shift of interest towards other coupling reactions.

One of the earliest examples of the use of Stille coupling in heterocyclic chemistry involves the conversion of *N*-methylpyrrole to 2-(trimethylstannyl)pyrrole in a lithiation, stannylation sequence, and its subsequent coupling with iodobenzene in the presence of tetrakis(triphenylphosphino)palladium (**6.27.**). The reaction proceeded smoothly in boiling THF to give the desired product in good yield.[36] A similar protocol involving lithium-halogen exchange and stannylation gave the corresponding 3-tributylstannylpyrrole, which was found to couple readily with aryl bromides.[37]

$$(6.27.)$$

Another entry into the preparation of 3-arylpyrroles starts with the reaction of the 3-iodopyrrole derivative shown in **6.28.** with hexabutyl-distannane in the presence of a palladium catalyst. The formed intermediate was reacted, in the presence of a similar catalyst system, with different aryl iodides to give the desired products in good to excellent yield.[38] It is worth mentioning that the presence of a formyl group in the 2-position of he pyrrole had no adverse effect on the efficiency of the couplings.

The Stille reaction was also effective in the coupling of two sensitive substrates. The functionalised dihydropyrrole and vinylstannane shown in **6.29.** were reacted under very mild conditions to give, after an acidic workup, the 2-acylpyrrole derivative.[39] The palladium catalyst contained the less strongly coordinating triphenylarsine ligand instead of triphenylphosphine, a "trick" commonly used to increase the efficiency of the Stille coupling.

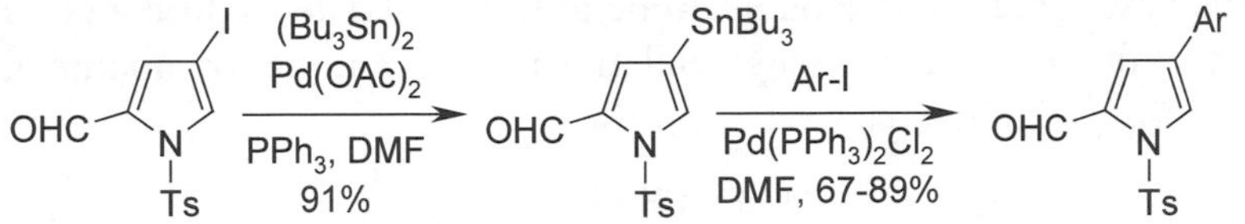

(6.28.)

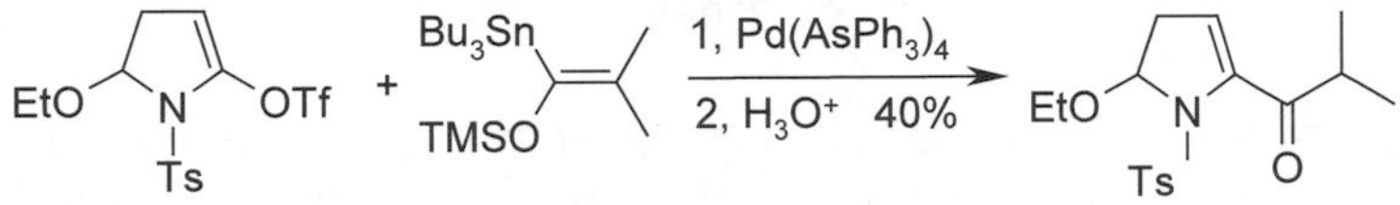

(6.29.)

The trialkylstannyl derivatives of furane are also frequently employed in Stille coupling. This reaction was utilised, for example, to introduce a chiral oxazoline moiety onto the furane core through the coupling of 2-trimethylstannylfurane and chiral 2-bromooxazoline derivatives (**6.30.**).[40] The furylstannane can be conveniently prepared in a lithiation-stannylation sequence, which makes it an attractive reagent for the introduction of the 2-furyl moiety.

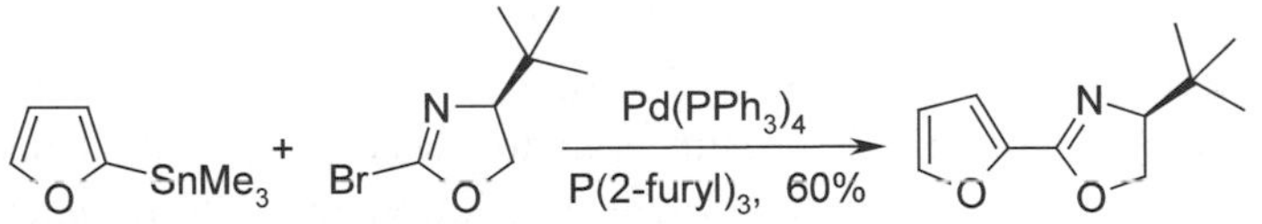

(6.30.)

In another example 2,5-bis(tributylstannyl)-furane was converted into 2,5-diarylfuranes in a double Stille coupling. Reaction with 4-bromo-benzonitrile, for example, give 2,5-bis(4'-cyanophenyl)-furane in good yield (**6.31.**).[41]

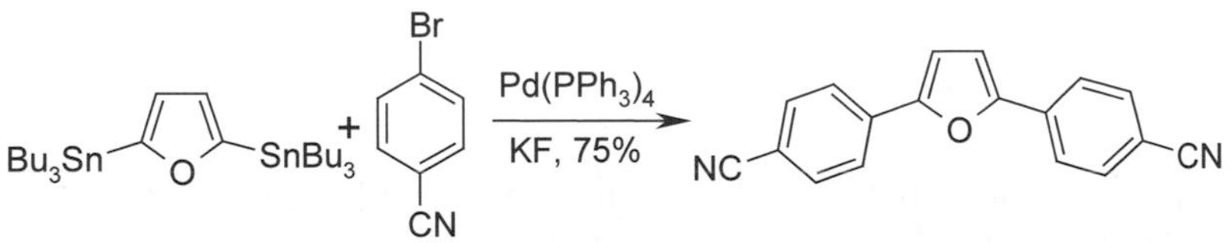

(6.31.)

2-(Trialkylstannyl)thiophenes are also easy to prepare and react in a manner similar to their furane analogues. 2-Tributylstannylfurane and 2-tributylstannylthiophene were both coupled, for example, with the 2-chloro-benzoxazole derivative shown in **6.32**. Both reactions proceeded efficiently to give the desired products in good yield.[42] The mild coupling conditions also allow for the use of polyfunctional coupling partners, a fact frequently exploited in medicinal chemistry.[43] The Stille coupling is also efficient in the preparation of benzothiophene derivatives, as has been demonstrated by Farina. The coupling was achieved using palladium-charcoal a catalyst source, triphenylarsine as ligand and copper(I)-iodide as co-catalyst.[44] The role of this latter additive is thought to be to transmetalate tin to copper and so facilitate the exchange (transmetalation) between the organometallic reagent and the palladium catalyst.[45]

(6.32.)

Copper(I) salts sometimes have a dual role in the Stille coupling. In the reaction of 3-methyltio-1,2,4-triazine and 2-tributylstannylthiophene (**6.33.**), for example, besides facilitating the transmetalation copper(I) is also crucial to activate the methyltio group to undergo oxidative addition.[46] The established Stille coupling conditions also worked well with other substrates, such as 2-methyltio-pyrimidine or 2-methyltio-benzthiazole.[47]

(6.33.)

Thienylstannanes were also coupled with a series of reagents other, than conventional aryl halides. 2,2'-Bithiophene was prepared from the thienyliodonium salt shown in **6.34.** and 2-tributylstannylthiophene. The activity of the iodonium salt allows for the room temperature coupling of the reagents in the presence of only 0.5 mol% palladium dichloride in a coordinating solvent mixture to give an excellent yield of the desired product.[48]

(6.34.)

The cephalosporine analogue shown in **6.35.** was also prepared in Stille coupling. Reaction of the allyl chloride moiety of chloromethylcephem and 2-tributylstannylthiophene, in the presence of a palladium-tris(2-furyl)phosphine catalyst, gave the coupled product in good yield without any interference from the reactive β-lactam moiety. The beneficial role of the furylphosphine derivative in the Stille coupling, analogously to triphenylarsine, stems from its weaker coordination to palladium compared with triphenylphosphine.[49]

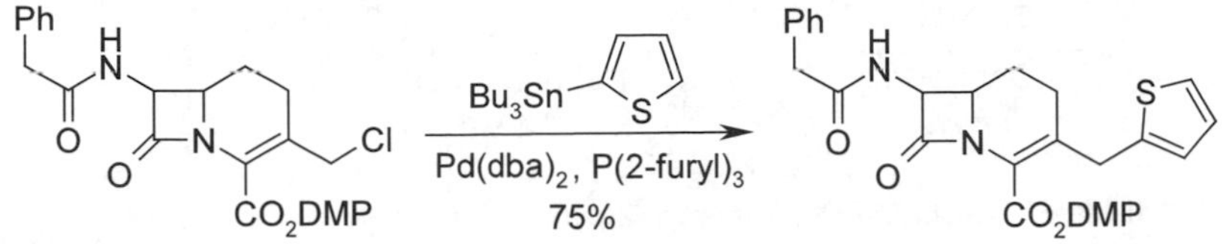

$$(6.35.)$$

The heterocyclic phosphate in **6.36.** was also successfully coupled with 2-tributylstannylthiophene. In this reaction the catalytic activity of the palladium-triphenylphosphine catalyst was boosted by the addition of lithium chloride.[50] The role of the additive, like in the coupling reactions of organic triflates, is to replace the weakly coordinating anion with chloride in the palladium complex formed in the oxidative addition step.

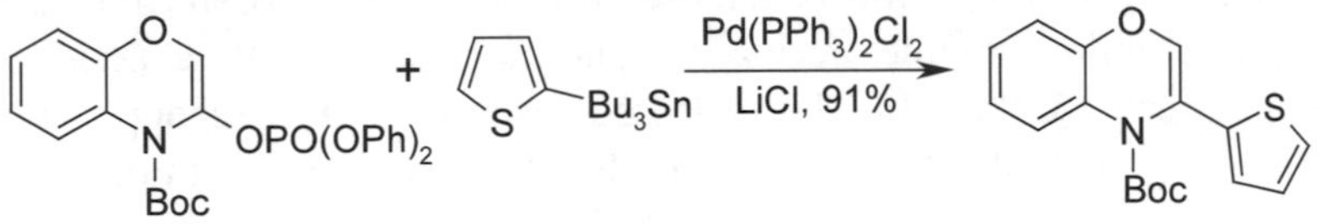

$$(6.36.)$$

The Stille coupling is regularly used for the functionalization of indole derivatives too. Indolylstannanes, either prepared in the reaction of lithioindoles with trialkyltin halides or by halogen-tin exchange with ditins, couple readily with aryl halides. 2-Tributylstannyl-tryptamine, prepared through the lithiation of tryptamine after the protection of its nitrogen atoms, was converted, for example, into a series of 2-aryl, vinyl and ethynyl derivatives in good yield (**6.37.**).[51] The analogous *N*-silylated 2-tributylstannylindole was equally efficient under the same coupling conditions.[52]

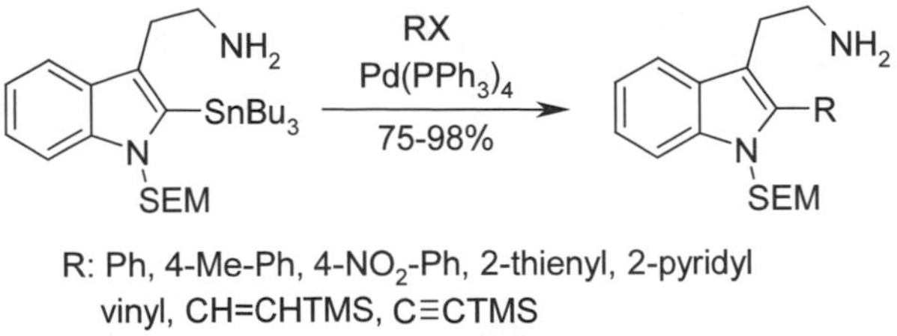

R: Ph, 4-Me-Ph, 4-NO$_2$-Ph, 2-thienyl, 2-pyridyl
vinyl, CH=CHTMS, C≡CTMS

$$(6.37.)$$

3-Tributylstannylindole derivatives were prepared from the appropriate 3-iodoindole through palladium catalyzed exchange with hexabutylditin (**6.38.**). The Stille coupling in the 3-position of the indole core, although less commonly used, is equally effective leading to the formation of a series of 3-

arylindoles in good yield using a palladium-triphenylarsine catalyst system. The reaction was exceptionally efficient with enol triflates.[53]

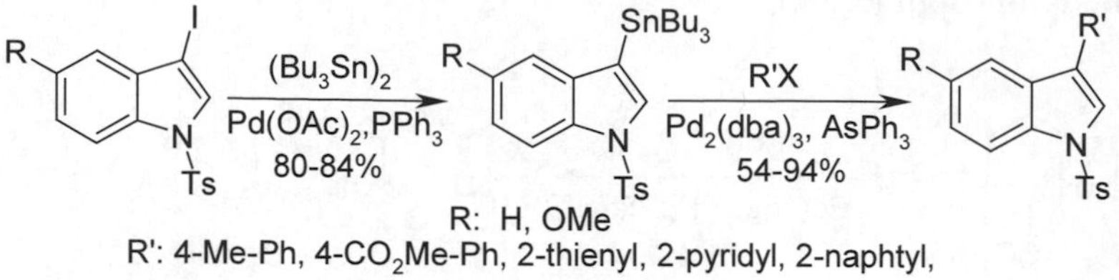

R: H, OMe
R': 4-Me-Ph, 4-CO$_2$Me-Ph, 2-thienyl, 2-pyridyl, 2-naphtyl,

(6.38.)

The Stille coupling of the organotin derivatives of imidazoles, oxazoles, thiazoles and their benzologues has also been investigated. *N*-methyl 2-tributylstannylimidazole, shown in **6.39.**, was coupled in the presence of a regular palladium catalyst with the 2,6-dibromo-isonicotinic acid ester to give 2,6-bis(imidazolyl)pyridine in good yield.[54] An analogous imidazolyltin compound was used efficiently to functionalise cytosine nucleosides.[55]

The functionalization of the isoxazole core was achieved starting from the appropriate iodoisoxazole derivative and treating it with a series of coupling partners (**6.40.**). In the selected examples aryl groups were introduced using both the Stille and the Suzuki coupling conditions. The transformations were extended to include the Heck and Suzuki couplings too. Attempts at the Kumada coupling, however, led to iodine-magnesium exchange instead of cross-coupling.[56]

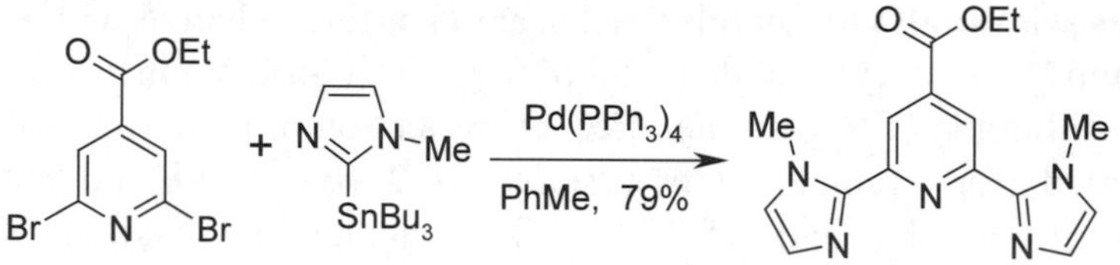

(6.39.)

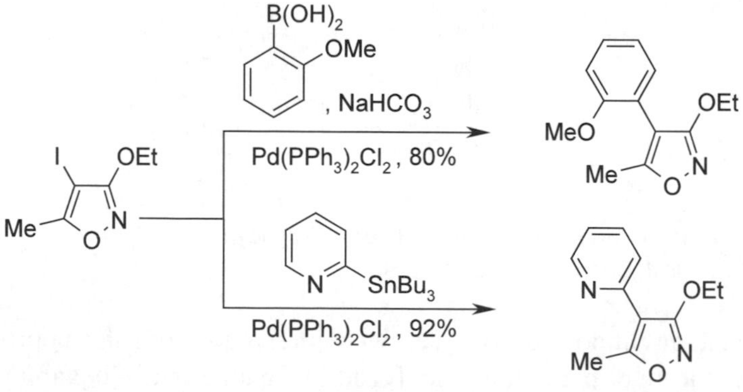

(6.40.)

2-Stannyloxazoles[57] and benzoxazoles are usually prepared through a lithiation-stannylation sequence as depicted for 2-tributylstannyl-benzoxazole in **6.41**.[58] 2-(Tributylstannyl)benzoxazole was coupled with bromobenzene to give 2-phenylbenzoxazole in good yield.[59] 2-Stannylthiazoles[60] and stannyl-benzothiazoles[57,61] were prepared analogously and reacted readily to give the 2-aryl derivatives.

$$(6.41.)$$

Finally, the outstanding functional group tolerance of the Stille reaction was exploited to prepare a series of γ-alkylidenebutenolids. γ-(Dibromomethylene)butenolide (**6.42.**) was sequentially coupled with phenyltributylstannane and styryltributylstannane to result in the selective exchange of the two bromine atoms. In the first intermediate it is always the Z-olefin that is formed.[62]

$$(6.42.)$$

Sonogashira coupling

The Sonogashira coupling of haloazoles and terminal acetylenes in the presence of a palladium(0)-copper(I) catalyst system usually proceeds readily. Its application has, in the beginning, been limited to iodoazoles, while recent examples frequently utilise bromo-heterocycles too.

Besides serving as a platform for the construction of the pyrrole ring (see Chapter 3.), the Sonogashira coupling is also effective in functionalising the same system. Trimethylsilylacetylene was used as a surrogate to introduce acetylene groups into the 2 and 5-positions of pyrrole (**6.43.**). 2,5-diiodo-1,3,4-trimethylpyrrole was reacted with the masked acetylene in the presence of the conventional tetrakis(triphenylphosphino)palladium catalyst and copper(I) iodide co-catalyst to give a mediocre yield of the desired product at room temperature. Removal of the protecting groups proceeded readily in the presence of base to give an excellent yield of the 2,5-diethynylpyrrole derivative.[63]

3-Iodopyrroles, such as the *N*-TIPS derivative shown in **6.44.** couple equally efficiently. Its reaction with a series of acetylene derivatives in the presence of a regular catalyst gave rise to the *N*-protected 3-ethynylpyrrole derivatives, which were converted to the appropriate pyrroles using tetrabutylammonium fluoride.[64]

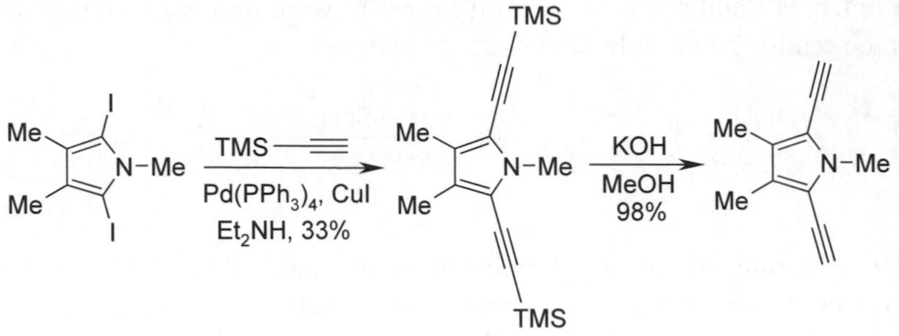

(6.43.)

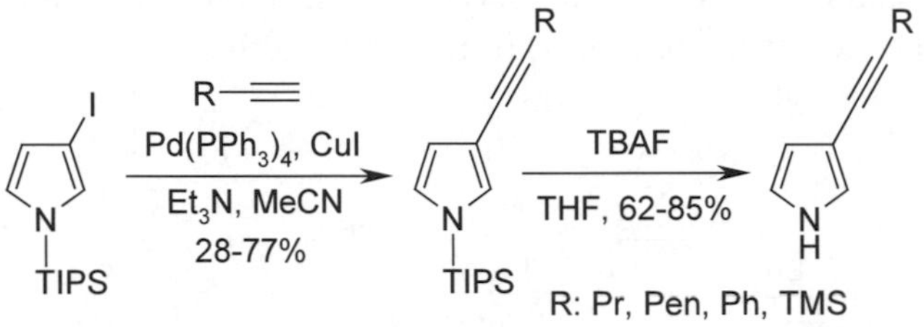

R: Pr, Pen, Ph, TMS

(6.44.)

Furane derivatives, like pyrroles, couple effectively with acetylenes. In case both the α-, and β-position are available for the reaction, like in the case of 4,5-dibromo-furfural in **6.45.**, the cross-coupling takes place preferentially in the former position.[65] The observed selectivity is in line with other palladium catalyzed transformations of dihalofuranes, such as their Stille coupling.[66]

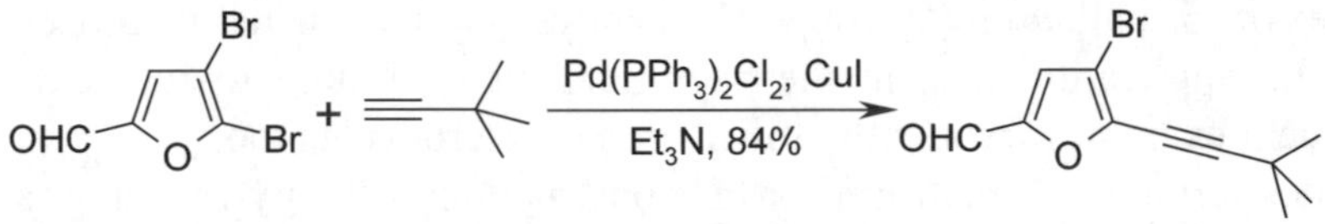

(6.45.)

The scope of the coupling of furane derivatives with acetylens was extended to alkynyltrifluoroborates too. These compounds are attracting increased attention due to their ready availability, chemical stability and willingness to participate in cross-coupling reactions. The furane derived bromo compound shown in **6.46.** for example reacted readily with a series of alkynyltrifluoroborates in the presence of a palladium-dppf catalyst to give the coupled products in excellent yield.[67]

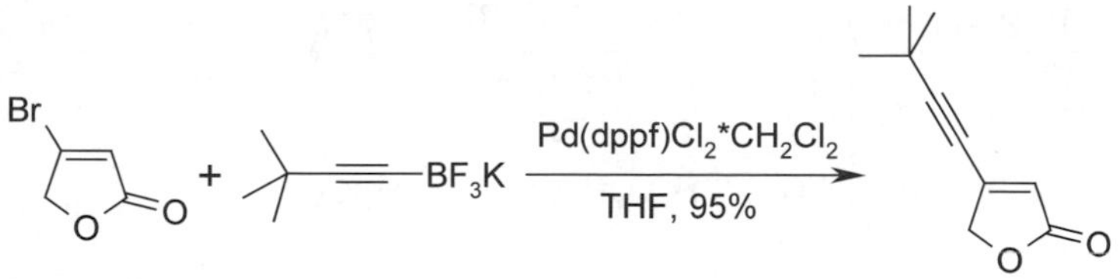

$$(6.46.)$$

The Sonogashira reaction of 2-iodothiophene with 2-methyl-3-butyne-2-ol or trimethylsilylacetylene under phase transfer conditions using sodium hydroxide as base led to the formation of the expected products, which released their end group spontaneously under the applied conditions giving rise to the intermediate formation of 2-ethynylthiophene. This terminal acetylene, in turn, reacted with another molecule of aryl halide, yielding either non symmetrical or symmetrical diarylethynes. When 2-methyl-3-butyn-2-ol was used as acetylene equivalent[68] it was possible to introduce a benzothiophene moiety in the second step, while the reaction of 2-iodothiophene and trimethylsilylacetylene led to the formation of 1,2-bis(2'-thienyl)acetylene (**6.47.**).[69]

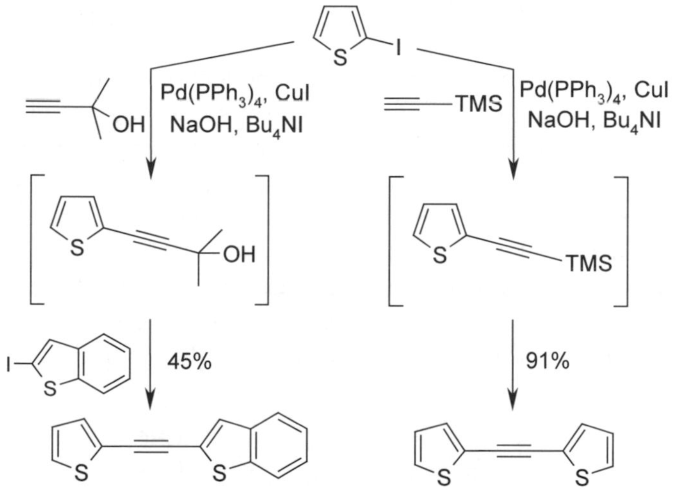

$$(6.47.)$$

An alternate approach to the palladium catalyzed ethynylation of thiophene derivatives has been reported by Zeni and co-workers. They coupled 2-(butyltelluro)thiophenes with different acetylenes (**6.48.**). The reaction, which was effectively catalyzed by palladium dichloride, was run in the absence of copper salts. The choice of base (triethylamine) and solvent (methanol) were both crucial for the success of the coupling.[70]

R: ^{n}Hex, CH_2OH, $(CH_2)_3OH$, $C(CH_3)(C_2H_5)OH$

(6.48.)

The benzologues of five membered heterocycles do also readily undergo Sonogashira coupling. 2,3-Dibromo-benzofuranes for example, were reacted with a series of alkynes and organozinc reagents to give selective coupling, analogously to furans, in the 2-position. The same selectivity was observed in the reactions of 2,3,5-tribromo-benzofurane, with no apparent coupling at positions 3 or 5 (6.49.).[71] Functionalization of the 3-position in the benzofurane ring was reported by Mann and co-workers, who reacted benzofuran-3-yl triflate under a succession of cross-coupling conditions to isolate the desired benzofurane derivatives in good yield.[72]

(6.49.)

Although in indole chemistry the primary application of the Sonogashira coupling is in the preparation of indole derivatives, the functionalization of the indole nucleus in positions 2 and 3 has also been accomplished using this methodology. Following the pioneering works of Yamanaka,[73] Gribble[74] and Prikhod'ko[75] a series of 2, and 3-alkynylindoles were prepared. A recent example, published by Larock, starts from the *N*-protected 2-bromo-3-formylindole and couples it with 1-decyne under standard conditions to give the expected product in 82% yield (6.50.). The analogous reaction of 2-formyl-3-iodo-1-methylindole and 3-butynol gave the coupled product in 87% yield.[76]

(6.50.)

Five membered heterocycles, containing more than 1 heteroatom were also used in Sonogashira reactions. 4-Iodopyrazole, protected in the form of its ethyl vinyl ether adduct (6.51.) was reacted with a series of acetylenes and the acidic workup of the crude product led to 4-ethynylpyrazole derivatives in good yield.[77]

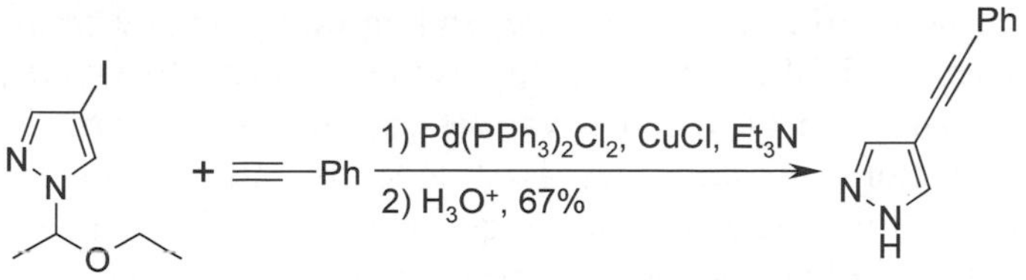

$$(6.51.)$$

Biologically active compounds, such as imidazolyl-ribofuranosides and hystidine derivatives,[78] were both prepared efficiently by the Sonogashira coupling. The protected ribofuranosylimidazole derivative shown in **6.52.** was coupled with 3-butynol in the absence of copper co-catalyst to give, after removal of the protecting groups, the desired ribosyl-heterocycle in 80% yield. It is interesting to note, that the same coupling gave only mediocre results when the conventional copper co-catalyst was also present.[79]

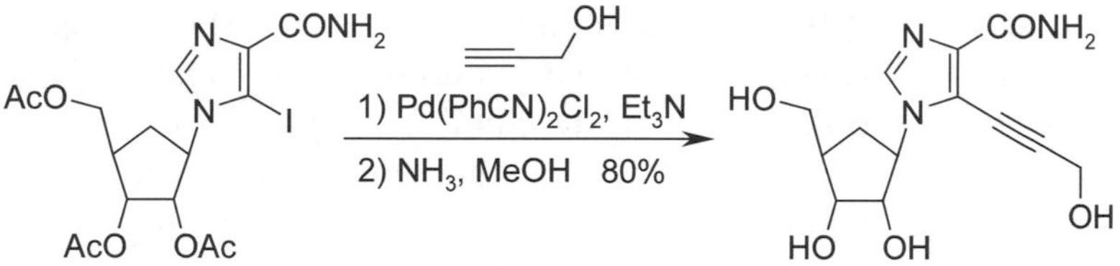

$$(6.52.)$$

The halogenated derivatives of oxazoles, thiazoles[80] and benzothiazoles[81] were also the subject of palladium catalyzed coupling with terminal acetylenes. 5-Bromo-2-methyl-4-phenyloxazole for example coupled efficiently with phenylacetylene using a conventional catalyst system to give an excellent yield of the desired product (**6.53.**).[82]

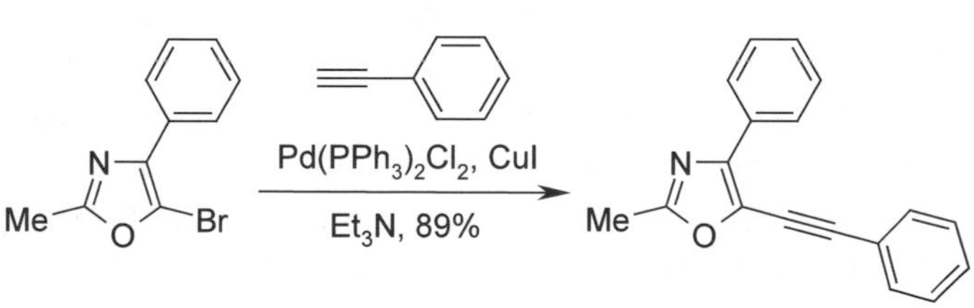

$$(6.53.)$$

6.2   INSERTION ROUTE

Azoles, like other heteroycles, usually undergo coupling reactions that involve the incorporation of an olefin or carbon monoxide readily. The insertion of carbon monoxide commonly leads to the formation of either a carboxylic acid derivative or a ketone, depending on the nature of the other reactants present. Unlike other heterocycles, most five membered heteroaromatics and their dihydro derivatives can also act as formal olefins in insertion reactions, a fact that is frequently utilised in synthetic transformations. With respect to this unique reactivity this chapter covers only the reactions of non-aromatic compounds, while the so called "heteroaryl Heck reactions" will be discussed in Chapter 6.4.

Heck reaction (olefin insertion)

As mentioned previously, the partially reduced forms of five membered heteroaromatic systems might act as olefins in insertion reactions. This behaviour is characteristic particularly of dihydrofuranes. The olefin insertion and the following β-hydride elimination should in principle lead to a trisubstituted olefin, which is rarely observed, however. Typical products of this reaction are 2-aryl-2,3-dihydrofuranes. A characteristic example of such a reaction is presented in **6.54**. The coupling of 4-iodoanisole and dihydrofurane led to the formation of the chiral 2-anisyl-2,3-dihydrofurane in excellent yield.[83] The shift of the double bond, which leads to the creation of a new centre of chirality in the molecule, opens up the way for enantioselective transformations. Both intermolecular and intramolecular variants of the asymmetric Heck reaction have been studied extensively.[84]

$$(6.54.)$$

The enantioselective introduction of aromatic substituents into the 3-position of the heterocycles is also possible, starting from 2,5-dihydroaromatics. In a representative example *N*-protected 2,5-dihydropyrrole was coupled with α-naphthyl tiflate in the presence of the chiral palladium-BINAP catalyst to give the 3-naphthyl-2,3-dihydropyrole derivative in moderate yield and enantioselectivity (**6.55.**).[85]

$$(6.55.)$$

The participation of halopyrroles in Heck coupling is mostly limited to intramolecular transformations. In a recent example of intermolecular Heck reaction different *N*-protected 3-iodo-4-trimethylsilyl-pyrroles were coupled with acrylate deivatives. In a typical example the Heck coupling with methyl acrylate under conventional conditions led to the formation of the desired product in 75% yield (**6.56.**).[86]

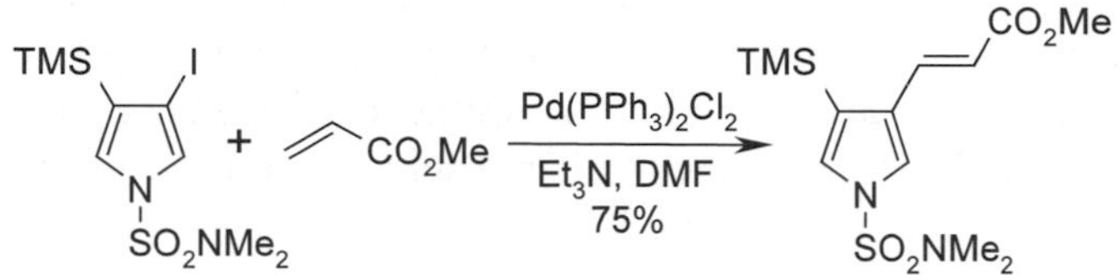

(6.56.)

Haloindoles and indolyl triflates are also known to undergo Heck coupling both in the 2- and 3-position. In a typical example the protected 3-indolyl triflate was coupled with ethyl acrylate in excellent yield (**6.57.**).[87]

The Heck reaction of a 2-bromoindole derivative was used to introduce a three-carbon chain onto the heterocyclic core *en route* to indoloquinolizidine alkaloids (**6.58.**).[88]

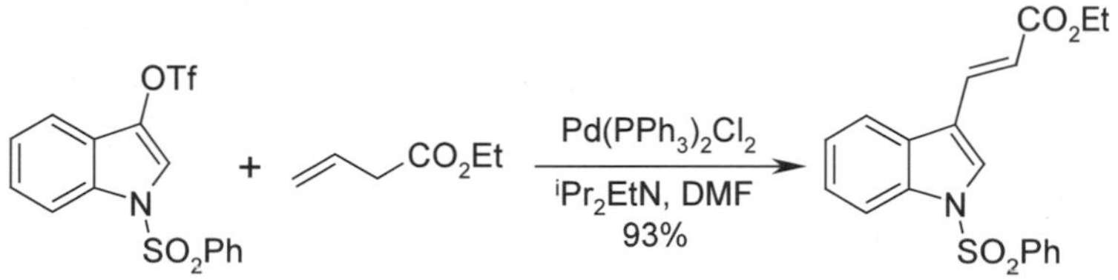

(6.57.)

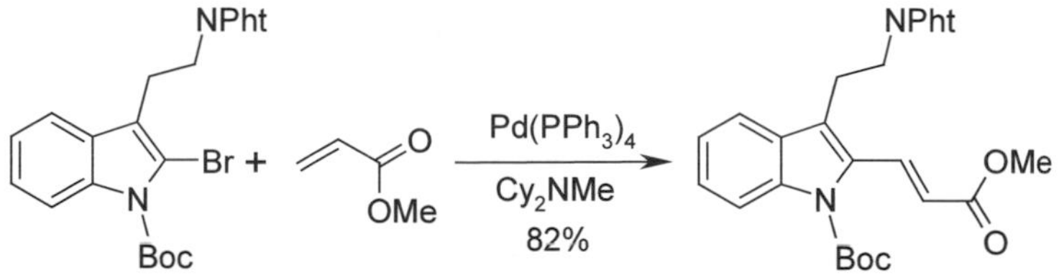

(6.58.)

The Heck reactions of thiophenes, particularly in the 2-position are well documented. In a recent example 2-bromothiophene was converted into a thienyl-vinylboronic acid derivative using a conventional palladium-triphenylphosphine catalyst and tributylamine as base (**6.59.**).[89]

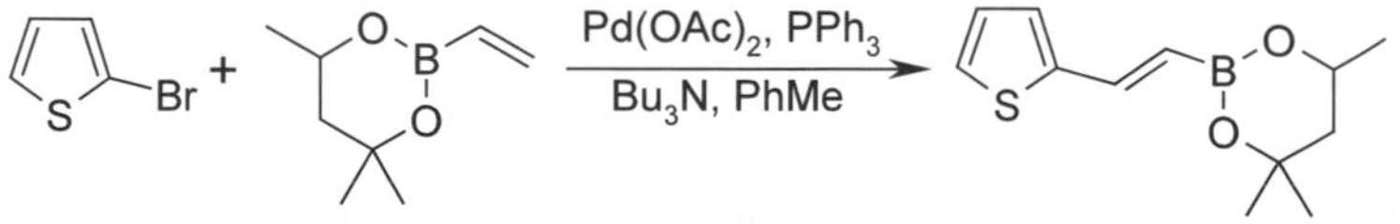

(6.59.)

The intermolecular Heck reactions of oxazoles and thiazoles with olefins are not too common. They are rarely high yielding since in several cases they are biased by dehalogenation. Due to this reason the olefination of these systems is usually achieved through Stille coupling with vinylstannanes.

An example of the Heck coupling of oxazols is presented in **6.60**. The 5-bromooxazole derivative reacted with styrene under standard conditions to give the expected styryloxazole in 87% yield. Coupling of the regioisomeric 4-bromooxazole gave 64%, while coupling of the more electron deficient acrylonitrile with the 5-bromooxazole gave only 50% of the desired product under the same conditions. The Heck coupling of the analogous thiazole derivative with ethyl acrylate gave similar results, leading to the formation of the expected product in 61% yield.[90]

(6.60.)

Insertion of an olefin can sometimes be accompanied by another coupling reaction. In the three component coupling of 2-iodothiophene, 1,1-dimethylallene and phenylboronic acid the insertion of the sterically less congested allene double bond leads to the formation of an allylpalladium complex, which undergoes transmetalation with phenylboronic acid and reductive elimination to give the product depicted in **6.61**. Since the Suzuki coupling of iodothiopene and phenylboronic acid is a potential side reaction, the reaction conditions, including solvent and base, had to be optimised carefully. The best selectivity and yield were obtained using $Pd(dba)_2$ as catalyst, CsF as base and DMF as solvent.[91]

(6.61.)

CO-insertion (carbonylative coupling)

The insertion of carbon monoxide into azolylpalladium complexes proceeds readily and in most cases leads to the formation of carboxylic acid derivatives or ketones. In a modified version of the carbonylation 3-bromothiophene was reacted with carbon monoxide in the presence of sodium formate. This reagents converts the intermediate acylpalladium formate complex, through the release of carbon dioxide into the acylpalladium hydride (*c.f.* **7.47.**), which in turn releases thiophene carboxaldehyde as the sole product (**6.62.**).[92] If sodium formate was replaced by potassium fluoride, then the appropriate thiophenecarboxylic acid fluoride was isolated from the reaction mixture.[93]

$$\text{(6.62.)}$$

The ready insertion of carbon monoxide into furanylpalladium complexes is impressively demonstrated by the reaction depicted in **6.63**. The iodofurane derivative was reacted with carbon monoxide in the presence of tetrabutylammonium chloride. Following an aqueous workup the appropriate carboxylic acid was isolated in good yield (**6.63.**).[94] It is worth pointing out, that due to the mildness of the reaction conditions the Heck coupling of the olefin moiety could be excluded.

$$\text{(6.63.)}$$

The carbonylation of thiazoles was achieved only using more forcing conditions. The prolonged heating of 5-bromothiazole in ethanol under CO pressure led to the formation of the desired ester in excellent yield (**6.64.**).[95]

Not only halothiophenes do undergo carbonylative coupling. The thienyliodonium salt, shown in **6.65.**, underwent carbon monoxide insertion followed by a coupling with anisylboronic acid under very mild conditions. Under an ambient atmosphere of carbon monoxide the reaction was complete in 1 hour at room temperature, and the desired anisylthiophene was isolated in 83% yield.[96] An analogous coupling utilised 2-(ethyldifluorosilyl)-thiophene.[97]

$$\text{(6.64.)}$$

$$\text{(6.65.)}$$

The insertion of carbon monoxide was also the key step in the formation of urea derivatives through the oxidative carbonylation of amines. 6-Amino-5-methylcarbazole was reacted with morpholine under an atmosphere of carbon monoxide and oxygen using palladium(II) iodide as catalyst to give the carbazolylurea derivative in good yield (**6.66.**).[98]

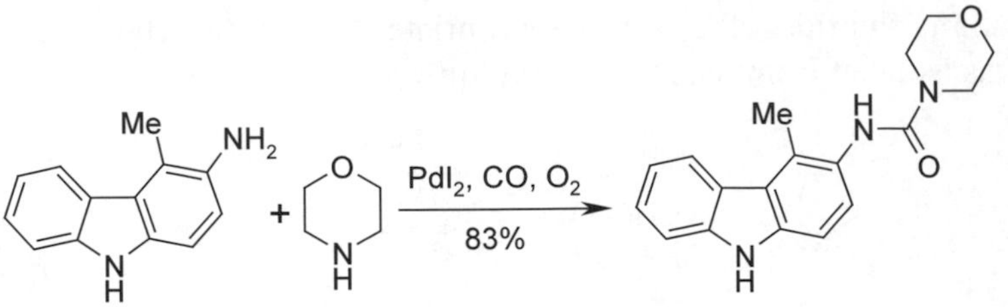

(6.66.)

6.3   CARBON-HETEROATOM BOND FORMATION

The introduction of nucleophiles onto five membered heterocycles through non-catalyzed aromatic nucleophilic substitution is of little synthetic value, since the comparatively high electron density of the aromatic ring makes the nucleophilic attack unfavourable. The introduction of transition metal catalyzed carbon-heteroatom bond forming reactions overcame this difficulty and led to a rapid increase in the number of such transformations.

A unique feature of azoles amongst five membered heterocycles is that they can act both as the carbon or the heteroatom donor during the bond formation. This possibility is frequently exploited in synthetic transformations. Pyrrole, for example, coupled effectively with bromoarenes in the presence of palladium based catalysts (6.67.). The use of P^tBu_3 as ligand and rubidium carbonate as base allowed for the reduction of catalyst loading to 1% without significant deterioration of the yield.[99]

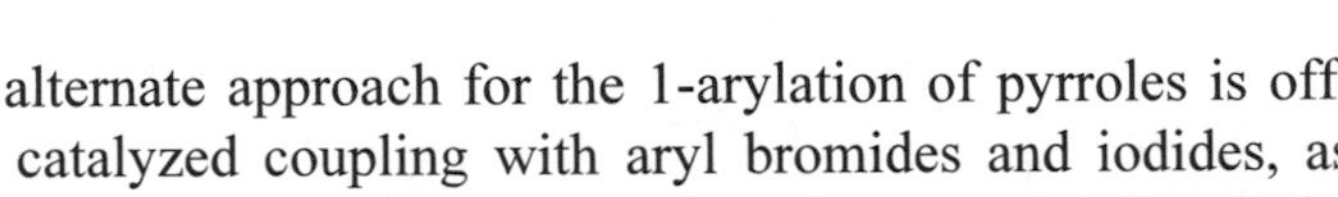

R	Reaction conditions	Yield, %
^tBu	6 mol% dppf, 5 mol% Pd(OAc)$_2$, NaO^{t}Bu, 120°C	87
H	3 mol% P^{t}Bu$_3$, 3 mol% Pd(dba)$_2$, Cs$_2$CO$_3$, 100°C	77
Me	3 mol% P^{t}Bu$_3$, 1 mol% Pd(OAc)$_2$, Rb$_2$CO$_3$, 120°C	70

(6.67.)

An alternate approach for the 1-arylation of pyrroles is offered by their copper catalyzed coupling with aryl bromides and iodides, as depicted in 6.68. The process, run in the presence of *trans-N,N'*-dimethyl-cyclohexane-diamine, was also efficient in the *N*-arylation of pyrazole, imidazole, triazoles and indazole, giving a mixture of isomers in the latter case.[100]

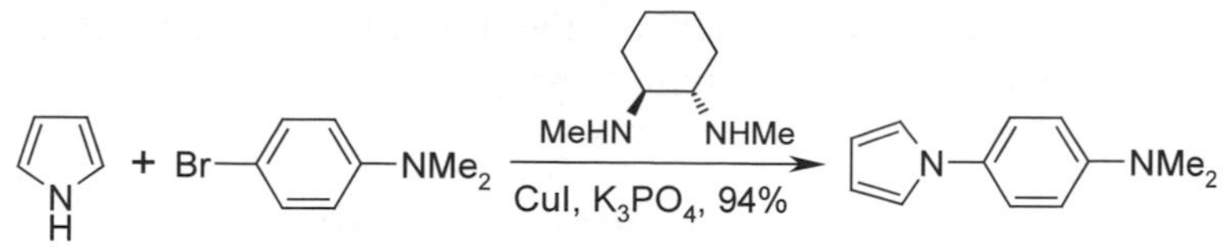

$$(6.68.)$$

The introduction of an imidazole moiety onto imidazo[1,2-*a*]pyridine was attempted using the same catalytic conditions as in **6.68**. Starting from 6-iodo-imidazopyridine the desired product was isolated in acceptable yield (55%), while the analogous 6-bromo compound gave only a disappointing 18% yield. Changing the solvent to the more polar DMF and the base to caesium carbonate led to an increase in the coupling efficiency giving 31% of the 6-imidazolyl derivative along with 37% of the 5-imidazolyl derivative arising through a *cine* substitution pathway (**6.69.**). By omitting the copper catalyst from the system the cine substitution becomes predominant, giving the 5-imidazolyl product in 58% yield. Other azoles showed a similar reactivity pattern.[101]

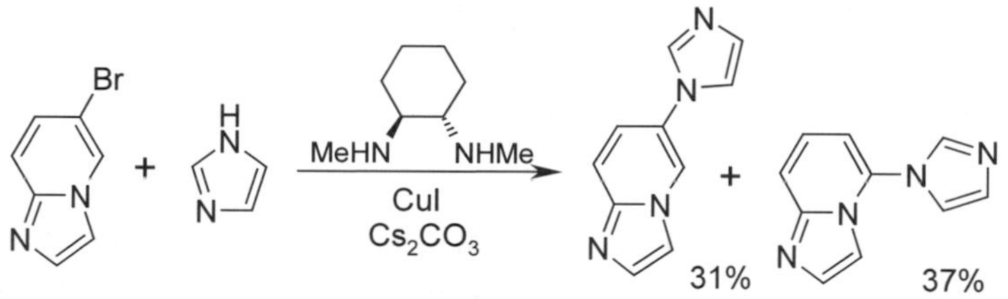

$$(6.69.)$$

Under copper catalyzed conditions azoles (*i.e.* imidazoles) couple not only with aryl halides but also with arylboronic acids. The reaction, run in the presence of oxygen, follows a unique path (for details see Chapter 2.5.). From the synthetic point of view, the arylation of imidazole proceeds in good yield, although the regioselectivity in the arylation of 4-substituted imidazoles is only moderate (**6.70.**).[102]

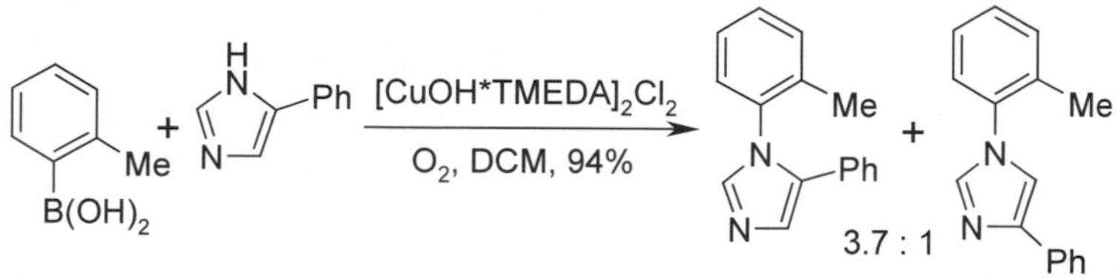

$$(6.70.)$$

The *N*-arylation of indoles has been achieved both in palladium and copper catalyzed coupling reactions. In an early example Hartwig succeeded in converting indole to 1-arylindoles in excellent yield, using a palladium-dppf catalyst and caesium carbonate as base (**6.71.**).[99a] Buchwald[99b] and

Watanabe[99c] improved the efficiency of the procedure by changing the ligand to P^{t}Bu$_3$ and the base to rubidium carbonate.

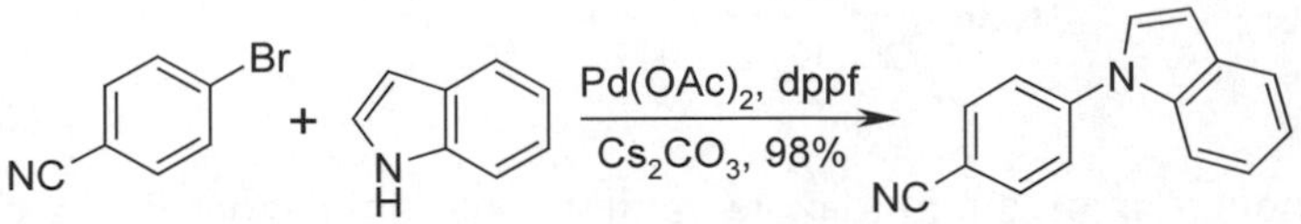

$$(6.71.)$$

Copper(I) iodide and *trans-N,N'*-dimethyl-cyclohexanediamine was also effective in converting indole derivatives into 1-arylindoles. The reaction of tryptamine and iodobenzene gave, for example, 1-phenyltryptamine selectively in 90% yield (**6.72.**). The selectivity of the process is highlighted by the coupling of indole and 2-iodoaniline, giving 1-(2'-aminophenyl)-indole in 88% yield.[103]

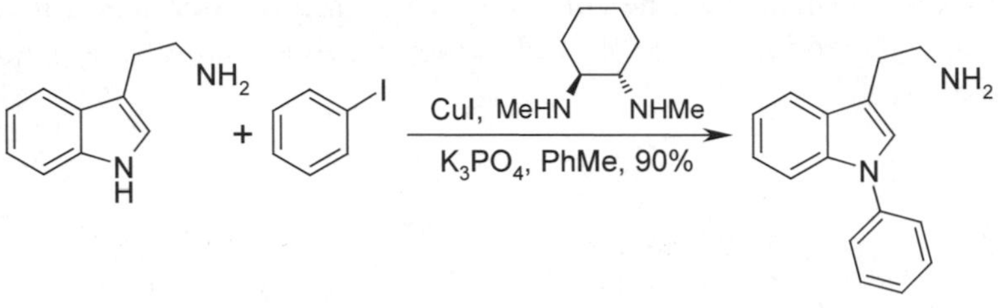

$$(6.72.)$$

Palladium and copper catalyzed carbon-hetroatom bond forming processes offer a complementary set of tools for the functionalization of heterocycles. 5-Aminoindole was arylated, depending on the catalyst used, either in the 1-position, or on the 5-amino group (**6.73.**). Its coupling with 3,5-dimethyl-iodobenzene in the presence of a copper-diamine catalyst gave the 1-arylindole derivative in 98% yield with over 20:1 selectivity. Alternately, the coupling of the same aminoindole and 3,5-dimethylphenol-benzenesulfonate in the presence of palladium and a sterically demanding biphenyl based ligand led to the arylation of the 5-amino group in 74% yield with 5:1 selectivity.[104]

In certain cases copper catalyzed processes might also be used for the *N*-alkynylation of azoles. Methyl indole-3-carboxylate was coupled with 1-bromo-2-triisopropylsilyl-acetylene in the presence of a copper-phenantroline catalyst to give the desired 1-ethynylindole derivative in excellent yield (**6.74.**).[105]

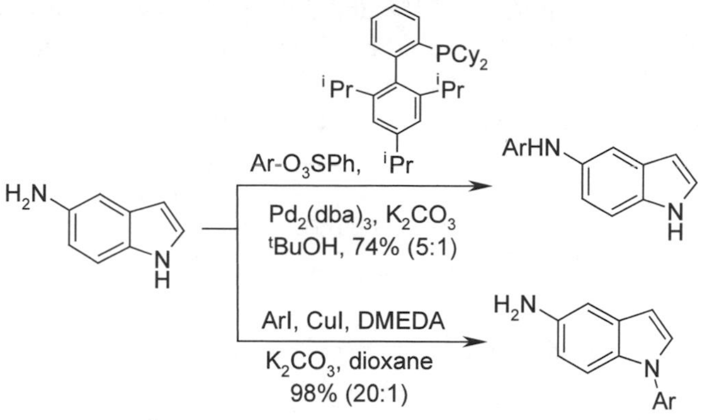

(6.73.)

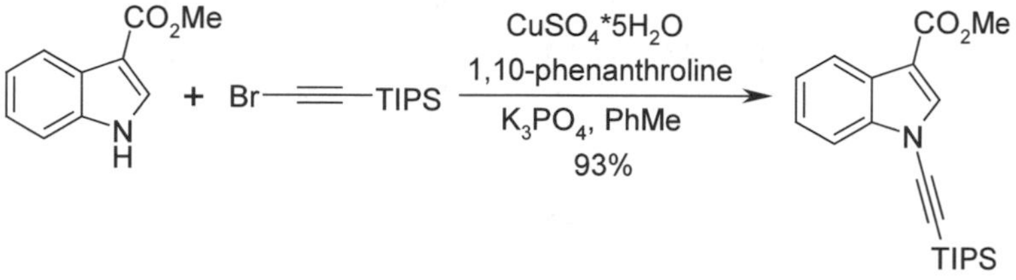

(6.74.)

Carbon-heteroatom bond forming reactions are also efficient in introducing amines onto other five membered heterocycles. 2-, and 3-bromothiophene were both coupled with diphenylamine using the highly active palladium-P^{t}Bu$_3$ catalyst system. The reactions furnished the desired products in both cases, although the yield varied significantly with the substitution pattern (**6.75.**).[106]

Certain 3-halothiophenes were successfully coupled with a series of amines in the presence of a palladium-BINAP catalyst system. The proper choice of the substituents on the thiophene ring as well as the choice of the halide were crucial for the success of the process (**6.76.**). 3-Bromo-2-thiophenecarboxylic acid derivatives, for example, coupled readily, while the conversion of 4-bromo-2-cyanothiophene failed completely.[107]

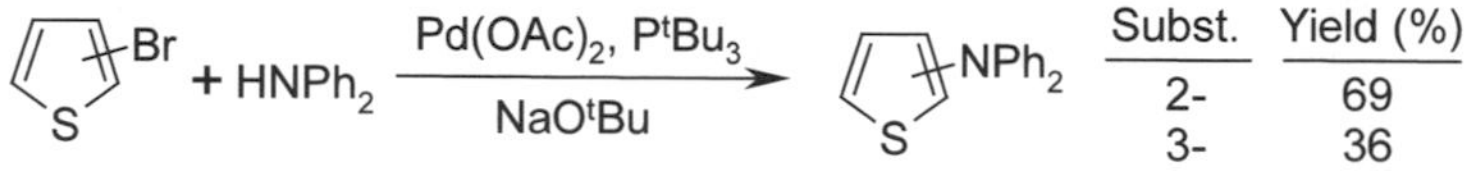

Subst.	Yield (%)
2-	69
3-	36

(6.75.)

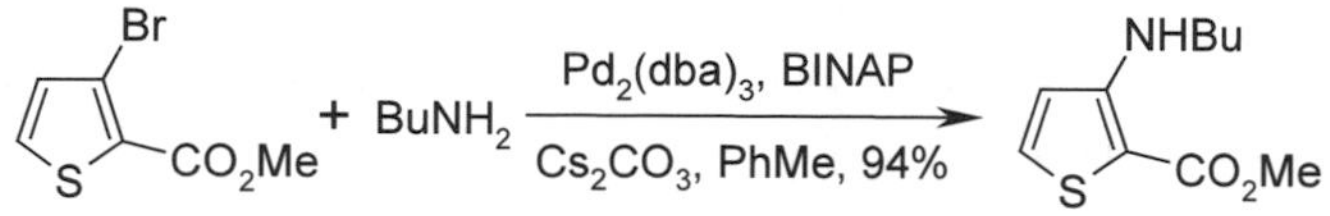

(6.76.)

Halofuranes also couple with amines, although usually less efficiently than halothiophenes. 2-, and 3-bromofurane were converted into the

appropriate anilino derivatives in 51% and 55% yield respectively, using a
P^{t}Bu$_3$-based catalyst (**6.77.**). The same conditions were also efficient for the
amination of other heterocyclic systems such as thiazole, indole,
benzothiophene, benzothiazole, benzoxazole and benzimidazole.[108]

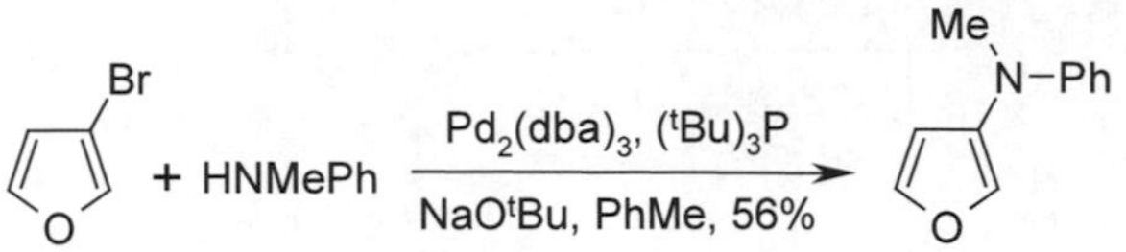

(6.77.)

The palladium based catalyst systems were also effective in the
construction of carbon-nitrogen bonds on benzannulated five membered
heterocycles. The 4-chloroindole derivative shown in **6.78.** was coupled with
piperazine in excellent yield, using a ferrocene based ligand.[109] The
analogous 5-bromo-benzimidazole derivative gave similar results (**6.79.**).[110]

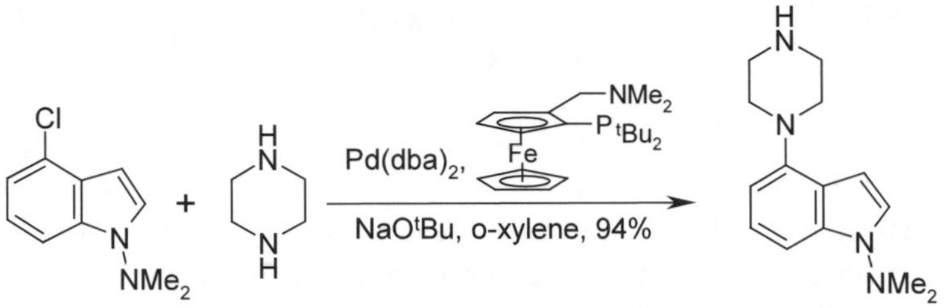

(6.78.)

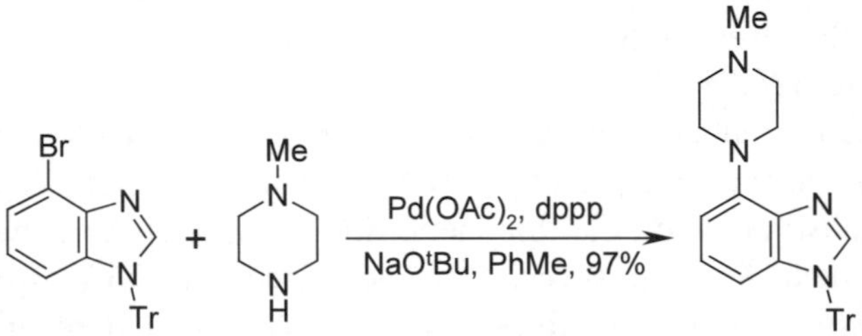

(6.79.)

In the copper catalyzed aromatic nucleophilic substitution of aryl halides
bromoindole derivatives were converted to the appropriate cyanoindoles.
Both 5-bromoindole and its *N*-tosyl derivative gave excellent yields, when a
substoichiometric amount potassium iodide was added to the reaction
mixture (**6.80.**). Pyrazole and benzothiophene showed a similar reactivity.
The role of the added iodide is to activate the aromatic system through a
bromine-iodine exchange.[111]

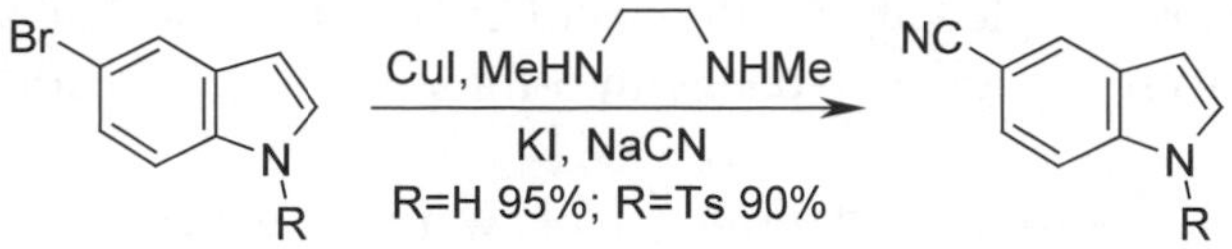

(6.80.)

Benzimidazoles bearing a halogen in the activated 2-position show a remarkable reactivity in palladium catalyzed carbon-nitrogen bond forming reactions. *N*-protected 2-chlorobenzimidazoles reacted smoothly with a series of amines (**6.81.**). The activity of the aryl halide, besides the ready coupling of the chloro derivative, is also emphasized by the low catalyst loading used.[112]

It was also demonstrated, that under similar coupling conditions related heterocycles, such as benzoxazoles and benzothiazoles react with a similar efficiency (**6.82.**). The coupling of piperidine with the above compound gave the 2-piperidino heterocycles in 66% and 78% yield, while the benzimidazole derivative gave 71%.[112]

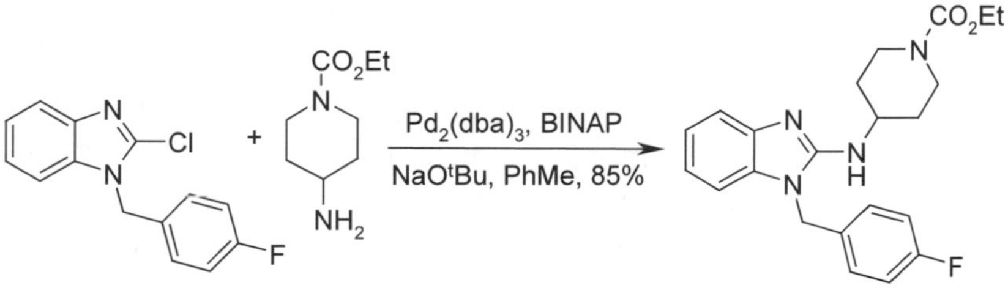

(6.81.)

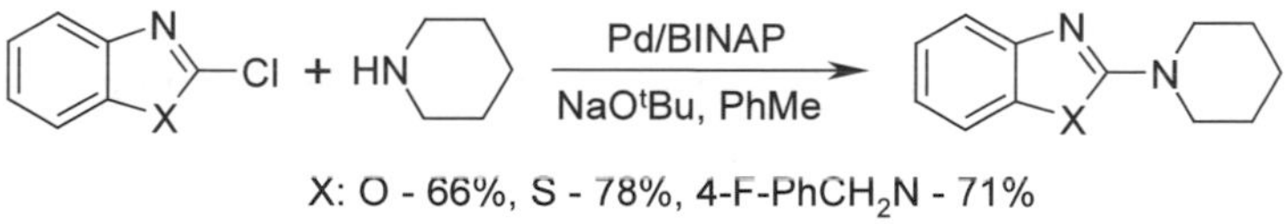

X: O - 66%, S - 78%, 4-F-PhCH$_2$N - 71%

(6.82.)

Examples of the transition metal catalyzed formation of carbon-heteroatom bonds other than carbon-nitrogen are less abundant. In a recent survey of the copper catalyzed carbon-oxygen bond formation between alcohols and organotrifluroborates the coupling of 3-thienyltrifluoroborate and 2-furfuryl alcohol gave the expected thienyl-furfuryl-ether in good yield (**6.83.**).[113]

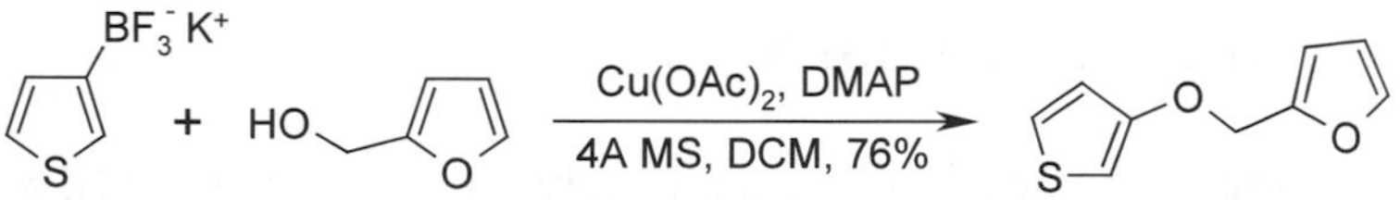

(6.83.)

There are two distinctly different copper catalyzed procedures that allow for the introduction of sulphur nucleophiles onto the thiophene core. 2-

Iodothiophene was coupled with thiophenol in the presence of a copper-neocuproin catalyst to give 2-phenylthio-thiophene in excellent yield. (**6.84.**).[114]

In the alternate approach arylboronic acids were coupled with *N*-arylthio-succinimides. In the process the aryl group of the boronic acid replaces the succinimide moiety in the coupling partner. In the example presented (**6.85.**) the 2-thiothiophene derivative was converted to 2-(4'-fluorophenylthio)-thiophene in 76% yield. The procedure was successfully extended to benzothiazoles too.[115]

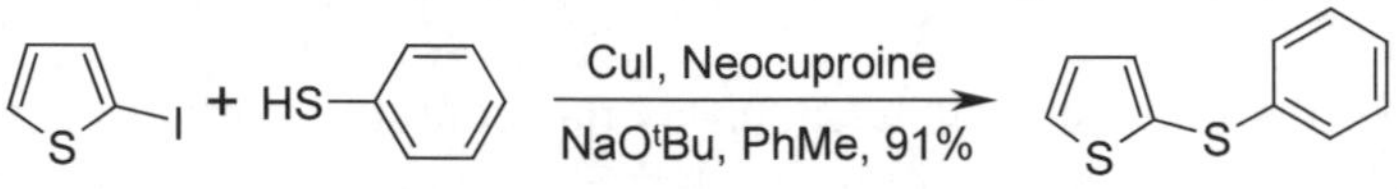

CuI, Neocuproine
NaO^{t}Bu, PhMe, 91%

(6.84.)

B(OH)$_2$
CuMeSal
76%

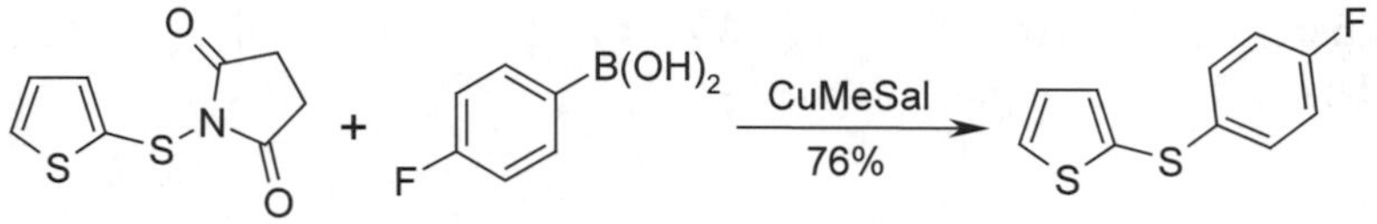

(6.85.)

Although not fitting exactly into the scope of this book, the iridium catalyzed borylation of five membered heterocycles through C-H bond activation also deserves mentioning. A recent report by Miyaura disclosed the reaction of bis(pinacolato)diboron with heteroaromatic systems, where thiophene, furane and pyrrole were converted to their 2-boryl derivatives with good selectivity (**6.86.**). The yields presented refer to the diboron compound since the heterocycles were used in excess in all cases. Indole, benzofurane and benzothiophene were monoborylated with similar efficiency.[116]

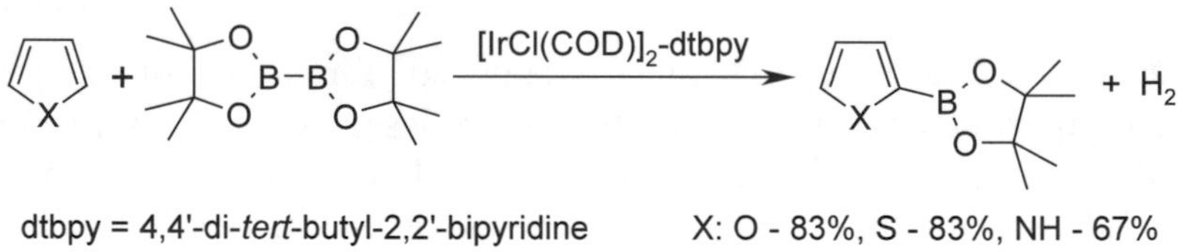

[IrCl(COD)]$_2$-dtbpy
+ H$_2$
dtbpy = 4,4'-di-*tert*-butyl-2,2'-bipyridine
X: O - 83%, S - 83%, NH - 67%

(6.86.)

6.4 OTHER PROCESSES

Most reactions discussed in this chapter rely on the formal attack of an electrophilic organometallic species on the electron-rich aromatic core of a five membered heterocycle. Depending on the way the transition metal complex is formed, these reactions are usually coined either "heteroaryl Heck reaction" or "C-H activation".[117]

The common feature of the first set of examples discussed is the coupling of an arylpalladium complex, formed in oxidative addition, with a five membered heterocyclic ring via the formal displacement of a hydrogen atom. This reaction, formally a Heck coupling, is often called the "heteroaryl Heck reaction".

N-Methylimidazole, when heated in DMF with bromobenzene in the presence of a palladium-triphenylphosphine catalyst and potassium carbonate furnished two products (**6.87.**): the 5-phenyl and the 2,5-diphenyl derivative. The product distribution suggests that the preferential site of the arylation is the more electron-rich 5-position.[118] Prolonged heating in a polar, high boiling solvent in the presence of base is characteristic of such transformations.

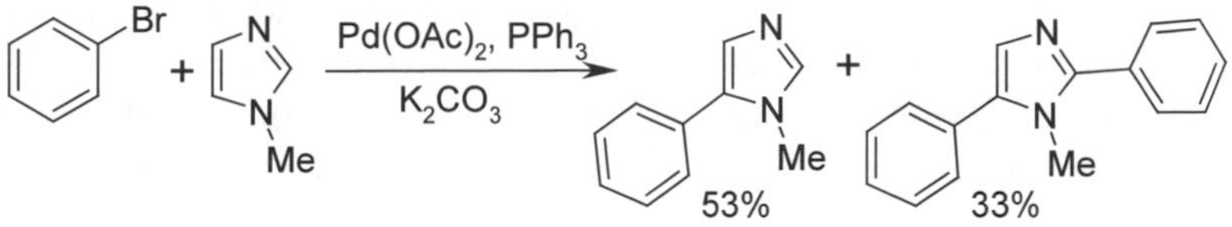

$$(6.87.)$$

Depending on the *N*-substituent, indole was arylated selectively in the 2- or 3-position. Starting from *N*-methylindole Ohta and co-workers obtained the 2-pyrazinyl-indole derivative in good yield (**6.88.**).[119] Increase of the steric bulk of the *N*-substituent diverts the incoming aryl moiety into the 3-position.

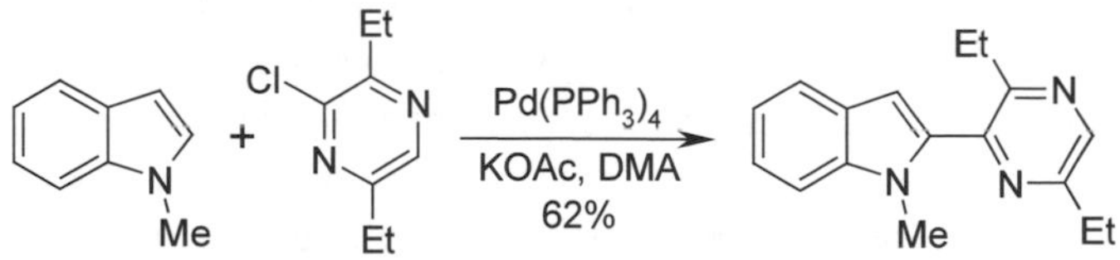

$$(6.88.)$$

The selective arylation of azoles, including imidazole, was achieved by Sames. Under the developed conditions, including the use of magnesium oxide as base, imidazole was arylated in the 4-position selectively, while the addition of a stoichiometric amount of copper(I) iodide led to the reversal of the regiochemistry, resulting in the selective formation of 2-phenylimidazole (**6.89.**).[120]

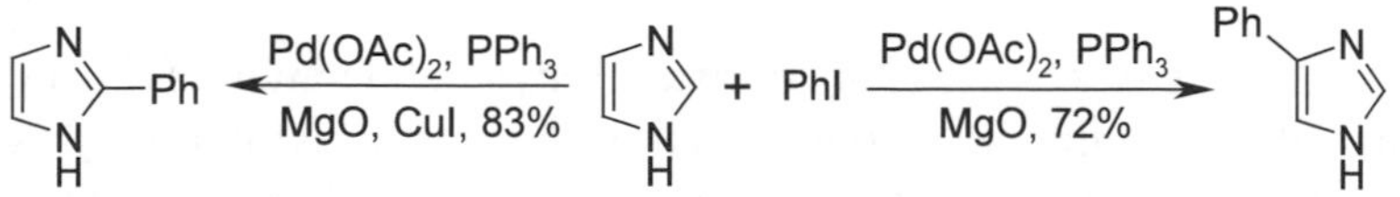

$$(6.89.)$$

Indolizines were arylated under similar conditions selectively in the 3-position (**6.90.**). A detailed mechanistic study of the transformation revealed that in this reaction the arylpalladium species, formed in the first step of the catalytic cycle, is attached to the indolizine core in an electrophilic substitution step, which is followed by reductive elimination. The presence of alternate routes such as Heck-type insertion, oxidative addition of the C-H bond, or transmetalation were excluded on the basis of experimental evidence.[121]

$$\text{indolizine} + \text{Br}-\text{thiophene} \xrightarrow[\text{KOAc, 55\%}]{\text{Pd(PPh}_3)_2\text{Cl}_2} \text{product}$$

(6.90.)

The use of the heteroaryl Heck reactions extends beyond fine chemicals synthesis. Polythiophenes were prepared starting from 3-octyl-2-iodotiophene by heating in the presence of palladium acetate and tetrabutylammonium chloride (**6.91.**).[122] The arylation of benzothiophene has also been achieved under the same conditions.[123]

$$\text{3-octyl-2-iodothiophene} \xrightarrow[\substack{\text{K}_2\text{CO}_3\text{, DMF, 93\%} \\ M_n = 3071,\ M_w = 6422}]{\text{Pd(OAc)}_2\text{, Bu}_4\text{NCl}} \text{polymer}$$

(6.91.)

Electron-rich heterocycles can also be coupled with olefins in the presence of a suitable palladium(II) catalyst. The oxidative coupling requires the use of a stoichiometric amount of palladium however, unless a suitable oxidising agent is added to the reaction. In an early example *N*-sulphonylated pyrrole was reacted with 1,4-naphthoquinone in the presence of an equimolar amount of palladium acetate to give the coupled product in good yield (**6.92.**).[124]

$$\text{N-SO}_2\text{Ph pyrrole} + \text{1,4-naphthoquinone} \xrightarrow[\text{AcOH, 55\%}]{\text{Pd(OAc)}_2} \text{product}$$

(6.92.)

In the analogous reaction of *N*-benzenesulfonyl-indole and ethyl acrylate the addition of an equimolar amount of silver(I) acetate as oxidising agent allowed for the lowering of the palladium content to catalytic levels. The coupling, proceeding selectively in the 3-position gave the indolylacrylate in excellent yield (**6.93.**).[125]

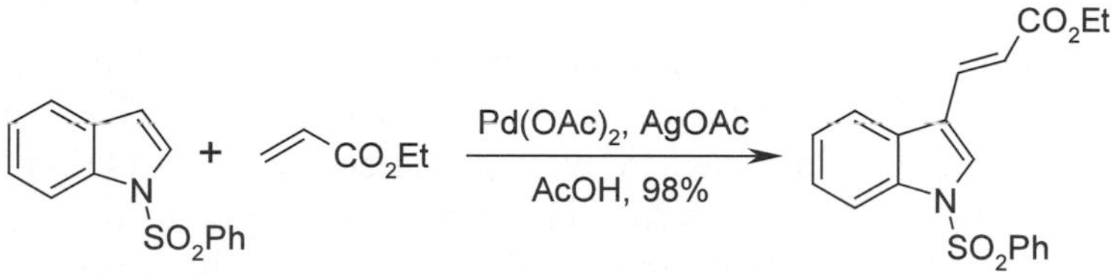

(6.93.)

The conversion of *N*-tosyl-4-bromoindole to the desired dehydrotryptophan derivative was achieved in high yield, using chloranil as the oxidising agent (**6.94.**). Screening experiments revealed that in this coupling chloranil is more effective than DDQ, Cu(OAc)$_2$, Co(salen)$_2$-O$_2$, or Ag$_2$CO$_3$.[126]

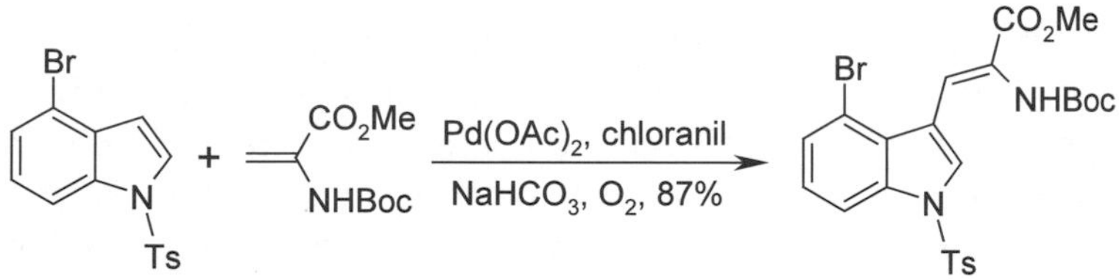

(6.94.)

By using an olefin embedded into the parent molecule Stoltz developed the oxidative annulation of indoles. The optimal catalyst consisted of palladium acetate and ethyl nicotinate, and molecular oxygen was used as the oxidant in the process. The reaction proceeded equally well irrespective of the attachment point of the alkyl chain bearing the pendant olefin bond on the five membered ring, and the formation of five and six membered rings were both effective (**6.95.**).[127]

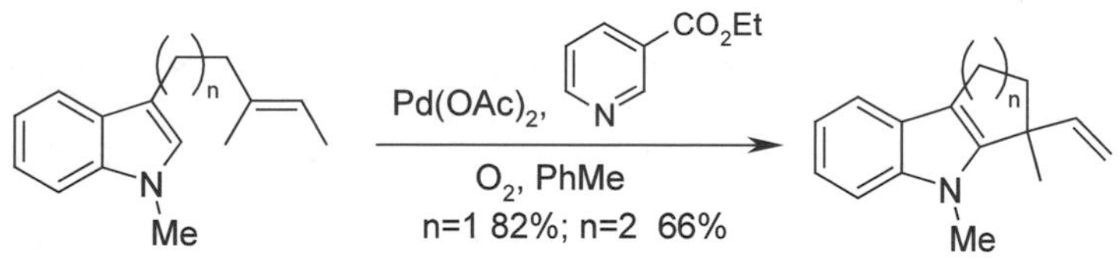

(6.95.)

Furans, thiophenes, thiazoles and pyrroles all reacted smoothly with alkylydenecyclopropanes in the presence of a palladium catalyst and added tributylphosphone, to give the allylated heterocycles as product. Thus furane-2-carboxylic ester on treatment with the butyl substituted alkylydenecyclopropane gave the 5-allyl-2-furanecarboxylate in good yield (**6.96.**).[128]

(6.96.)

In a unique reaction 2-iodothiophene was converted to 2-benzoylthiophene using benzaldehyde as reagent (**6.97.**). The nickel catalyzed reductive coupling, which was promoted by the addition of a stoichiometric amount of zinc as reducing agent, gave access to the product in mediocre yield (*N.B.* reversing the coupling roles gave the same product in 47% yield).[129]

(6.97.)

The copper catalyzed oxidative dimerisation of arylboronic acids was also used for the preparation of symmetrical bithiophene derivatives. 2-Formylthiophene-3-boronic acid and 5-boronic acid were both dimerised successfully giving the appropriate dithiophenes in 35% and 41% yield (**6.98.**). The optimised coupling conditions included running the reactions in DMF in the presence of 50 mol% copper(II) acetate.[130]

(6.98.)

6.5   REFERENCES

[1] (a) Miyaura, N.; Suzuki, A. *Chem. Rev.* **1995**, *95*, 2457. (b) Stanforth, S. S. *Tetrahedron* **1998**, *54*, 263. (c) Suzuki, A. *J. Organomet. Chem.* **1999**, *576*, 147.

[2] For a recent review on heterocyclic boronic acids see Tyrrel, E.; Brookes, P. *Synthesis*, **2003**, 469.

[3] Johnson, C. N.; Stemp, G.; Anand, N.; Stephen, S. C.; Gallagher, T. *Synlett* **1998**, 1025.

[4] Thoresen, L. H.; Kim, H.; Welch, M. B.; Burghart, A.; Burgess, K. *Synlett* **1998**, 1276.

[5] Routier, S.; Coudert, G.; Mérour, J.-Y. *Tetrahedron Lett.* **2001**, *42*, 7025.

[6] Nettekoven, M.; Püllmann, B.; Schmitt, S. *Synlett* **2003**, 1649.

[7] Haddach, M.; McCarthy, J. R. *Tetrahedron* **1999**, *40*, 3109.

[8] Sotgiu, G.; Zambianchi, M.; Barbarella, G.; Botta, C. *Tetrahedron* **2002**, *58*, 2245.

[9] Melucci, M.; Barbarella, G.; Sotgiu, G. *J. Org. Chem.* **2002**, *67*, 8877.

[10] Chambers, M. S.; Atack, J. R.; Broughton, H. B.; Collinson, N.; Cook, S.; Dawson, G. R.; Hobbs, S. C.; Marshall, G.; Maubach, K. A.; Pillai, G. V.; Reeve, A. J.; MacLeod, A. M. *J. Med. Chem.* **2003**, *46*, 2227.

[11] Malapel-Andrieu, B. Mérour, J.-Y. *Tetrahedron* **1998**, *54,* 11079.
[12] Jiang, W.; Sui, Z.; Macielag, M. J.; Walsh, S: P.; Fiordeliso, J. J:; Lanter, J. C.; Guan, J.; Qui, Y.; Kraft, P.; Bhattacharjee, S.; Craig, E:; Haynes-Johnson, D.; John, T. M.; Clancy, J. J. Med. Chem. **2003**, *46*, 441.
[13] Kumar, J. S. D.; Ho, MK. M.; L., J. M.; Toyokuni, T. *Adv. Synth. Catal.* **2002**, *344*, 1146.
[14] Conway, S. C.; Gribble, G. W. *Heterocycles* **1990**, *30*, 627.
[15] Dinnell, K.; Chicchi, G. G.; Dhar, M. J.; Elliott, J. M.; Hollingworth, G. J:; Kurtz, M. M.; Ridgill, M. P.; Rycroft, W.; Tsao, K.-L.; Williams, A. R.; Swain, C. J. *Bioorg. & Med. Chem. Lett.* **2001**, *11,* 1237.
[16] Kawasaki, I.; Yamashita, M.; Ohta, S. *Chem. Pharm. Bull.* **1996**, *44*, 1831.
[17] Nguyen, H. N.; Huang, X.; Buchwald, S. L. *J. Am. Chem. Soc.* **2003**, *125*, 11818.
[18] Liebeskind, L. S.; Srogl, J. *Org. Lett.* **2002**, *4*, 979.
[19] Yang, C.-G.; Liu, G.; Jiang, B. *J. Org. Chem.* **2002**, *67,* 9392.
[20] Bumagin, N. A.; Nikitina, A. F.; Beletskaya, I. P. *Russ J. Org. Chem.* **1994**, *30*, 1619.
[21] Minato, A.; Tamao, K.; Hayashi, T.; Suzuki, K.; Kumada, M. *Tetrahedron Lett.* **1981**, *22*, 5319.
[22] An alternate approach to these compound classes consists of the iron mediated oxidative coupling of organometallic reagents. For an example see Lukevics, E.; Arsenyan, P.; Pudova, O. *Heterocycles* **2003**, *60*, 663.
[23] Abotto, A.; Bradamante, S.; Facchetti, A.; Pagani, G. A. *J. Org. Chem.* **1997**, *62*, 5755.
[24] Lucas, L. N.; De Jong, J. J. D.; van Esch, J. H.; Kellogg, R. M.; Feringa, B. L. *Eur J. Org. Chem.* **2003**, 155.
[25] Xia, C.; Fan, X.; Locklin, J.; Advincula, R. C. *Org. Lett.* **2002**, *4*, 2067.
[26] Brandão, M. A. F.; de Oliveira, A. B.; Snieckus, V. *Tetrahedron Lett.* **1993**, *34,* 2437.
[27] Takahashi, K.; Gunji, A. *Heterocycles* **1996**, *43,* 941.
[28] Miyasaka, M.; Rajca, A. *Synlett* **2004** 177.
[29] Bach, T.; Heuser, S. *J. Org. Chem.* **2002**, *67,* 5789.
[30] Felding, J.; Kristensen, J.; Bjerregaard, T.; Sander, L.; Vedsø, P.; Begtrup, M. *J. Org. Chem.* **1999**, *64*, 4196.
[31] Vincent, P.; Beaucourt, J. P.; Pichart, L. *Tetrahedron Lett.* **1984**, *25*, 201.
[32] Amat, M.; Hadida, S.; Pshenichnyi, G.; Bosch, J. *J. Org. Chem.* **1997**, *62*, 3158 and references therein.
[33] Bell, A. S.; Roberts, D. A.; Ruddock, K. S. *Tetrahedron Lett.* **1988**, *29*, 5013.
[34] Massa, M. A.; Patt, W. C.; Ahn, K.; Sisneros, A. M.; Herman, S. B.; Doherty, A. *Bioorg. Med. Chem. Lett.* **1998**, *8*, 2117.
[35] Luo, F.-T.; Wang, R.-T. *Heterocycles* **1990**, *31*, 1543.
[36] Bailey, T. R. *Tetrahedron Lett.* **1986**, *27,* 4407.
[37] Alvarez, A.; Guzmán, A.; Ruiz, A.; Velarde, E.; Muchowski J. M. *J. Org. Chem.* **1992**, *57,* 1653.
[38] Wang, J.; Scott, A. I. *Tetrahedron Lett.* **1996**, *37*, 3247.
[39] Luker, T.; Hiemstra, H.; Speckamp, W. N. *Tetrahedron Lett.* **1996**, *37*, 8257.
[40] Meyers, A. I.; Novachek, K. A. *Tetrahedron Lett.* **1996**, *37*, 1747.
[41] Kumar, A.; Stephens, C. E.; Boykin, D. W. *Heterocyclic Commun.* **1999**, *5,* 301.
[42] Batista-Parra, A:; Venkitachalam, S.; Wilson, W. D.; Boykin, D. W. *Heterocycles* **2003**, *60*, 1367.
[43] Boschelli, D. H.; Wang, D. Y.; Ye, F.; Yamashita, A.; Zhang, N.; Powell, D.; Weber, J.; Boschelli, F. *Bioorg. Med. Chem. Lett.* **2002**, *12,* 2011.
[44] Roth, G. P.; Farina, V. *Tetrahedron Lett.* **1995**, *36*, 2191.

[45] (a) Farina, V.; Kapadia, S.; Krishnan, B.; Wang, C.; Liebeskind *J. Org. Chem.* **1994**, *59*, 5905. (b) Han, X.; Stoltz, B. M.; Corey, E.J. *J. Am. Chem. Soc.* **1999**, *121*, 7600.

[46] Alphonse, F.-A.; Suzenet, F.; Keromnes, A.; Lebret, B.; Guillaumet, G. *Org. Lett.* **2003**, *5*, 803.

[47] Egi, M.; Liebeskind, L. S. *Org. Lett.* **2003**, *5*, 801.

[48] Kang, S.-K.; Lee, H.-W.; Jang, S.-B.; Kim, T.-H.; Kim, J.-S. *Synth. Commun.* **1996**, *26*, 4311.

[49] Park, H.; Lee, J. Y.; Lee, Y. S.; Park, J. O.; Koh, S. B.; Ham, W-H. *J. Antibiotics* **1994**, *47*, 606.

[50] Buon, C.; Bouyssou, P.; Coudert, G. *Tetrahedron Lett.* **1999**, *40*, 701.

[51] Palmisano, G.; Santagostino, M. *Synlett* **1993**, 771.

[52] Palmisano, G.; Santagostino, M. *Helv. Chim. Acta* **1993**, *76*, 2356.

[53] Ciattini, P. G.; Morera, E.; Ortar, G. *Tetrahedron Lett.* **1994**, *35*, 2405.

[54] Vermonden, T.; Branowska, D.; Marcelis, A. T. M.; Sudhölter, E. J. R. *Tetrahedron* **2003**, *59*, 5039.

[55] Gutierrez, A. J.; Terhorst, T. J.; Matteucci, M. D.; Froehler, B. C. *J. Am. Chem. Soc.* **1994**, *116*, 5540.

[56] Kromann, H.; Sløk, F. A.; Johansen, T. N.; Krogsgaard-Larsen, P. *Tetrahedron*, **2001**, *57*, 2195.

[57] Dondoni, A.; Fantin, G.; Fogagnolo, M.; Medici, A.; Pedrini, P. *Synthesis* **1987**, 693.

[58] The use of trimethyltin chloride in place of the tributyltin derivative leads predominantly to the formation of a ring opened compound. Jutzi, P.; Gilge, U. *J. Organomet. Chem.* **1983**, *246*, 159.

[59] Kosugi, M.; Koshiba, M.; Atoh, A.; Sano, H.; Migita, T. *Bull. Chem. Soc. Jpn.* **1986**, *59*, 677.

[60] Dondoni, A.; Mastellari, A. R.; Medici, A.; Negrini, E. *Synthesis* **1986**, 757.

[61] Molloy, K. C.; Waterfield, P. C.; Mahon, M. F. *J. Organomet. Chem.* **1989**, *365*, 61.

[62] Sorg, A.; Siegel, K.; Brückner, R. *Synlett* **2004**, 311.

[63] Ortaggi, G.; Scarsella, M.; Scialis, R.; Sleiter, G. *Gazz. Chim. Ital.* **1988**, *118*, 743.

[64] Alvarez, A.; Guzmán, A.; Ruiz, A.; Velarde, E.; Muchowski J. M. *J. Org. Chem.* **1992**, *57*, 1653.

[65] Bach, T.; Krüger, L. *Eur. J. Org. Chem.* **1999**, 2045.

[66] Bach, T.; Krüger, L. *Synlett* **1998**, 1185.

[67] Kabalka, G. W.; Dong, G.; Venkataiah, B. *Tetrahedron Lett.*, **2004**, *45*, 5139.

[68] Carpita, A.; Lessi, A.; Rossi, R. *Synthesis* **1984**, 571.

[69] D'Auria, M. *Synth. Commun.* **1992**, *22*, 2393.

[70] (a) Zeni, G.; Nogueira, C. W.; Panatieri, R. B.; Silva, D. O.; Menezes, P. H.; Braga, A. L.; Silveira, C. C.; Stefani, H. A.; Rocha, J. B. T. Tetrahedron Lett. 2001, 42, 7921. (b) For a review of the coupling of tellurides see: Zeni, G.; Braga, A. L.; Stefani, H. A. *Acc. Chem. Res.* **2003**, *36*, 731.

[71] Bach, T.; Bartels, M. *Synthesis* **2003**, *6*, 925.

[72] Morice, C.; Garrido, F.; Mann, A.; Suffert, J. *Synlett* **2003**, *3*, 501.

[73] Sakamoto, T.; Nagano, T.; Kondo, Y.; Yamanaka, H. *Chem. Pharm. Bull.* **1988**, *36*, 2248.

[74] Gribble, G.; Conway, S. C. *Synth. Commun.* **1992**, *22*, 2129.

[75] Prikhod'ko, T. A.; Kurilenko, V. M.; Khlienko, Zh. N.; Vasilevskii, S. F.; Shvatrsberg, M. S. *Izv. Akad. Nauk SSSR, Ser. Khim.* **1990**, 134.

[76] Zhang, H.; Larock, R. C. *Tetrahedron Lett.* **2002**, *43*, 1359.

[77] Vasilevsky, S. F; Klyatskaya, S. V.; Tretyakov, E. V.; Elguero, J. *Heterocycles* **2003**, *60*, 879.

[78] Evans, D. A.; Bach, T. *Angew. Chem. Int. Ed.* **1993**, *32*, 1326.

[79] Minakawa, N.; Takeda, T.; Sasaki, T.; Matsuda, A.; Ueda T. *J. Med. Chem.* **1991**, *34*, 778.

[80] Sakamoto, T.; Nagata, H.; Kondo, Y.; Shiraiwa, M.; Yamanaka, H. *Chem, Pharm. Bull.* **1987**, *35*, 823.

[81] Schegel, J.; Maas, G. *Synthesys* **1999**, 100.

[82] Sagamoto, T.; Nagata, H.; Kondo, Y.; Shiraiwa, M.; Yamanaka, H. *Chem. Pharm. Bull.* **1987**, *35*, 823.

[83] Amengual, R.; Genin, E.; Michelet, V.; Savignac, M.; Genet, J.-P. *Adv. Synth. Catal.* **2002**, *344*, 393.

[84] For reviews on the asymmetric Heck reaction see: (a) Dounay, A. B.; Overmann, L. E. Chem. Rev. 2003, 103, 2945. (b) Shibasaki, M.; Miyazaki, F. in *Handbook of Organopalladium Reagents for Organic Synthesis*, Ed.: Negishi, E-i., Wiley, Hoboken, **2002**, Vol. 1, p 1283.

[85] Sonesson, C.; Larhed, M.; Nyqvist, C.; Hallberg, A. *J. Org. Chem.* **1996**, *61*, 4756.

[86] Liu, J.-H.; Chan, H.-W.; Wong, H. N. C. *J. Org. Chem.* **2000**, *65*, 3274.

[87] Gribble, G. W.; Conway, S. C. *Synth. Commun.* **1992**, *22*, 2129.

[88] Luo, S.; Zificsak, C. A.; Hsung, R. P. *Org. Lett.* **2003**, *5*, 4709.

[89] Lightfoot, A. P.; Maw, G.; Thirsh, C.; Twiddle, S. J. R.; Whiting, A. *Tetrahedron Lett.* **2003**, *44*, 7645.

[90] Sagamoto, T.; Nagata, H.; Kondo, Y.; Shiraiwa, M.; Yamanaka, H. *Chem. Pharm. Bull.* **1987**, *35*, 823.

[91] Chang, T.-H. H. H-M.; Wu, M.-Y.; Cheng, C.-H. *J. Org. Chem.* **2002**, *67*, 99.

[92] Okano, T.; Harada, N.; Kiji, J. *Bull. Chem. Soc. Jap.* **1994**, *67*, 2329.

[93] Okano, T.; Harada, N.; Kiji, J. *Bull. Chem. Soc. Jap.* **1992**, *65*, 1741.

[94] Gagnier, S. V.; Larock, R. C. *J. Am. Chem. Soc.* **2003**, *125*, 804.

[95] Head, R. A.; Ibbotson, A. *Tetrahedron Lett.* **1984**, *25*, 5939.

[96] Kang, S.-K.; Lim, K.-H.; Ho, P.-S.; Yoon, S.-k.; Son, H.-J. *Synth. Commun.* **1998**, *28*, 1481.

[97] Hatanaka, Y.; Fukushima, S.; Hiyama, T. *Tetrahedron* **1992**, *48*, 2113.

[98] Gabriele, B.; Salerno, G.; Mancuso, R.; Costa, M. *J. Org. Chem.* **2004**, *69*, 4741.

[99] (a) Mann, G.; Hartwig, J. F.; Driver, M. S.; Fernandez-Rivas, C. *J. Am. Chem. Soc.* **1998**, *120*, 827. (b) Hartwig, J. F.; Kawatsura, M.; Hauck, S. I.; Shaughnessy, K. H.; Alcazar-Roman, L. M. *J. Org. Chem.* **1999**, *64*, 5575. (c) Watanabe, M.; Nishiyama, M.; Yamamoto, T.; Koie, Y. *Tetrahedron Letters* **2000**, *41*, 481.

[100] Antilla, J. C.; Baskin, J. M.; Barder, T. E.; Buchwald, S. L. *J. Org. Chem* **2004**, *69*, 5578.

[101] Enguehard, C.; Allouchi, H.; Gueiffier, A.; Buchwald, S. L. *J. Org. Chem.* **2003**, *68*, 5614.

[102] Collman, J. P.; Zhong, M. *Org. Lett.* **2000**, *2*, 1233.

[103] Antilla, J. C.; Klapars, A.; Buchwald, S. L. *J. Am. Chem. Soc.* **2002**, *124*, 11685.

[104] Huang, X.; Anderson, K. W.; Zim, D.; Jiang, L.; Klapars, A.; Buchwald, S. L. *J. Am. Chem. Soc.* **2003**, *125*, 6653.

[105] Zhang, Y.; Hsung, R. P.; Tracey, M. R.; Kurtz, K. C. M.; Vera, E. L. *Org. Lett.* **2004**, *6*, 1151.

[106] Watanabe, M.; Yamamoto, T.; Nishiyama, M.; *Chem. Commun.* **2000**, 133.

[107] Luker, T. J.; Beaton, H. G.; Whiting, M.; Mete, A.; Chershire, D. R. *Tetrahedron Lett.* **2000**, *41*, 7731.

[108] Hooper, M. W.; Utsonomiya, M.; Hartwig, J. F. *J. Org. Chem.* **2003**, *68*, 2861.

[109] Watanabe, M.; Yamamoto, T.; Nishiyama, M. *Angew. Chem. Int. Ed.* **2000**, *39*, 2501.

[110] (a) Lópes-Rodríguez, M. L.; Viso, A.; Benhamú, B.; Romiguera, J. L.; Murica, M.; *Bioorg. Med. Chem. Lett.* **1999**, *9*, 2339. (b) Lópes-Rodríguez, M. L.; Benhamú, B.;

Ayala, D.; Romiguera, J. L.; Murica, M.; Ramos, J. A.; Viso, A. *Tetrahedron* **2000**, *56*, 3245.

[111] Zanon, J.; Klapars, A.; Buchwald, S. L. *J. Am. Chem. Soc.* **2003**, *125*, 2890.

[112] (a) Tanoury, G.J.; Senanayake, C.H.; Hett, R.; Kuhn, A.M.; Kessler, D.W.; Wald, S.A. *Tetrahedron* **1998**, *54*, 6845. (b) Hong, Y.; Senanayake, C.H.; Xiang, T.; Vandenbossche , C.P.; Tanoury, G.J.; Bakale, R.P.; Wald, S.A. *Tetrahedron Lett.* **1998**, *39*, 3121.

[113] Quach, T. D.; Batey, R. A. *Org. Lett.* **2003**, *5*, 1381.

[114] Bates, C. G.; Gujadhur, R. K.; Venkatamaran, D. *Org. Lett.* **2002**, *4*, 2803.

[115] Savarin, C.; Srogl, J.; Liebeskind, L. S. *Org. Lett.* **2002**, *4*, 4309.

[116] Takagi, J.; Sato, K.; Hartwig, J. F.; Ishiyama, T.; Miyaura, N. *Tetrahedron Lett.* **2002**, *43*, 5649.

[117] For a recent review on coupling reactions proceeding via C-H activation see Miura, M.; Nomura, M. in *Cross-coupling reactions*, *Topics in Chemistry Vol. 219.*, Miyaura, N. Ed., Springer Verlag, Beréin, Heidelberg, **2002**, pp 211-241.

[118] Pivsa-Art, S.; Satoh, T.; Kawamura, Y.; Miura, M.; Nomura, M. *Bull. Chem. Soc. Jpn.* **1998**, 71, 467.

[119] Akita, Y.; Igataki, Y.; Takizawa, S.; Ohta, A. *Chem. Pharm. Bull.* **1989**, 37, 1477.

[120] Sezen, B.; Sames, D. *J. Am. Chem. Soc.* **2003**, *125*, 5274.

[121] Park, C.-H.; Ryabova, V.; Seregin, I. V.; Sromek, A. W.; Gevorgyan, V. *Org. Lett.* **2004**, *6*, 1159.

[122] Sévignon, M.; Papillon, J.; Schulz, E.; Lemaire, M. *Tetrahedron Lett.* **1999**, *40*, 5873.

[123] Chabert, J. F. D.; Gozzi, C.; Lemaire, M. *Tetrahedron Lett.* **2002**, *43*, 1829.

[124] Itahara, T. *J. Org. Chem.* **1985**, *50*, 5546.

[125] Itahara, T.; Kawasaki, K.; Ouseto, F. *Synthesis* **1984**, 236.

[126] (a) Yokoyama, Y.; Matsumoto, T.; Murakami, Y. *J. Org. Chem.* **1995**, *60*, 1486. (b) Osanai, K.; Yokoyama, Y.; Kondo, K.; Murakami, Y. *Chem. Pharm. Bull.* **1999**, *47*, 1587.

[127] Ferreira, E. M.; Stoltz, B. M. *J. Am. Chem. Soc.* **2003**, *125*, 9578.

[128] Nakamura, I.; Siriwardana, A. I.; Satio, S.; Yamamoto, Y. *J. Org. Chem.* **2002**, *67*, 3445.

[129] Huang, Y.-C.; Majumdar, K. K.; Cheng, C.-H. *J. Org. Chem.* **2002**, *67*, 1682.

[130] Demir, A. S.; Reis, Ö.; Emrullahoglu, M. *J. Org. Chem.* **2003**, *68*, 10130.

Chapter 7

THE FUNCTIONALIZATION OF SIX MEMBERED RINGS

The transition metal catalyzed functionalization of six membered heterocycles constitutes a major chapter of this book. The abundance of examples might be tracked down to two reasons: *i*) the large number of available haloazines; *ii*) the marked reactivity of haloazines in catalytic reactions, which originates in their electron deficient nature. In general the reactivity of a 2-chloropyridine derivative is comparable to a bromoarene, while a bromopyridine behaves more like an iodobenzene derivative. The increasing number of nitrogen atoms in the ring makes chlopyridazines or chloropyrimidines reactive enough to participate in transition metal catalyzed processes. The potential biological activity of six membered heterocyclic systems and their abundance as structural motif in natural products drove synthetic chemists to exploit their transformations. This chapter is mainly dedicated to the reactions of monocyclic six membered heterocycles and their benzologues. The chemistry of other condensed systems of importance, such as purines and pyrons, is discussed in Chapter 8.

7.1 TRANSMETALATION ROUTE

Probably the most thoroughly studied transition metal catalyzed transformation of six membered systems is their participation in cross-coupling reactions. Due to the vastness of this field we present only some representative examples for the different reaction types. In this chapter reactions where a haloazine is used to introduce the six membered ring on the palladium in the oxidative addition and processes, where the azine moiety is transferred onto the catalyst in a transmetalation step will be discussed parallel. The featured reactions were divided into subclasses along the commonly used name reactions.

Suzuki coupling

The availability of reagents and their functional group tolerance led to the emergence of Suzuki coupling as the most utilized cross-coupling reaction of the recent years.[1] Either haloazines are coupled with an organoboron reagent, or the azines are converted into boronate esters, boronic acids or boranes and react with aryl halides, the reaction usually proceeds equally well. The examples presented below will focus more on the latter case, the coupling of hetarylboron compounds.[2] The coupling of pyrone derivatives is discussed in Chapter 8.1.

In equation **7.1.** a 4-chloropyridine was coupled with diethyl(3-pyridyl)borane.[3] The reaction was run in aqueous THF in the presence of potassium carbonate. The role of the base is to facilitate the transmetalation step through the formation of a borate ion, as organoboranes are usually not nucleophilic enough to transfer their organic moiety onto the palladium. An alternate function of the base is to increase the electrophilicity of the palladium through exchange of the halide to carbonate.

$$\text{Pd(PPh}_3)_4$$
$$\text{K}_2\text{CO}_3, \text{THF/H}_2\text{O}$$
$$65\%$$

(7.1.)

A similar transformation was reported using a polyfunctional phthalazine derivative (**7.2.**).[4] It is interesting to note that the transmetalation of the borane was not selective with respect to the pyridyl group, formation of the product resulting from the transfer of the ethyl group was also observed.

This side reaction is not observed in each case. Coupling of the same pyridylborane with 2-bromo-nitrobenzene (**7.3.**) led to the formation of 3-(o-bromophenyl)-pyridine in good yield.[5] In the reaction running in boiling THF potassium hydroxide was used as base and the addition of some phase transfer catalyst (TBAB) was found to increase the efficiency of the coupling.

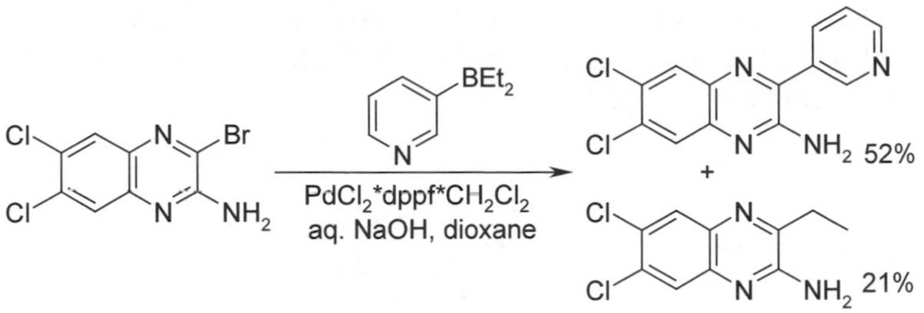

(7.2.)

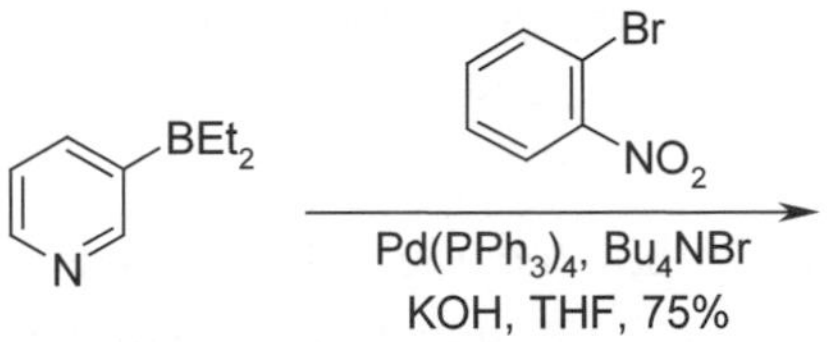

(7.3.)

A potential way to avoid the formation of undesired side products, like in **7.2.**, is the use of such boron compounds that have only one transferable group. In most cases boronic acids are the compounds of choice, as they are easy to prepare, insensitive to moisture and air, and usually form crystalline solids. In certain cases, however the transmetalation of the heteroaryl group might be hindered by the formation of stable hydrogen bonded complexes. In such cases the use of a boronate ester, such as in equation **7.4.**, provides better yields. For example pyridine-2-boronic acid dimethylester coupled readily with a bromoquinoline derivative under conditions similar to **7.3.** (potassium hydroxide was used as base and tetrabutylammonium bromide as phase transfer catalyst).[6]

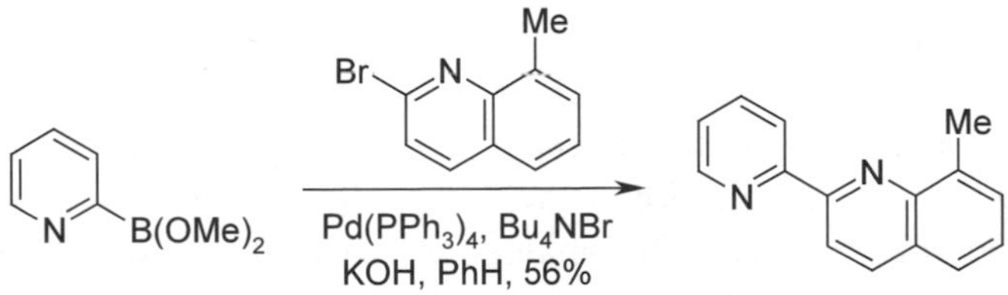

(7.4.)

The most common way for the preparation of pyridylboronic acids is through the reaction of lithiated pyridines with boronic esters. The lithiation of the pyridine core is usually achieved through lithium-halogen exchange from the corresponding bromopyridine or in the presence of a directing metalation group (DMG) through *ortho*-lithiation. The boronic acids shown in **7.5.** were prepared from 2,5-dihalopyridines with good selectivity (in 79% and 61% yield respectively) using *n*-butyllithium for the lithium-halogen exchange, (triisopropyl)borate for the introduction of the boron moiety and aqueous hydrolysis to form the boronic acid. The coupling of the chloro derivative and 3-bromoquinoline proceeded readily to furnish the desired product in 55% yield, while the bromo derivative gave the analogous product in 32%. The coupling was carried out in both cases in aqueous DMF using sodium carbonate as base.[7]

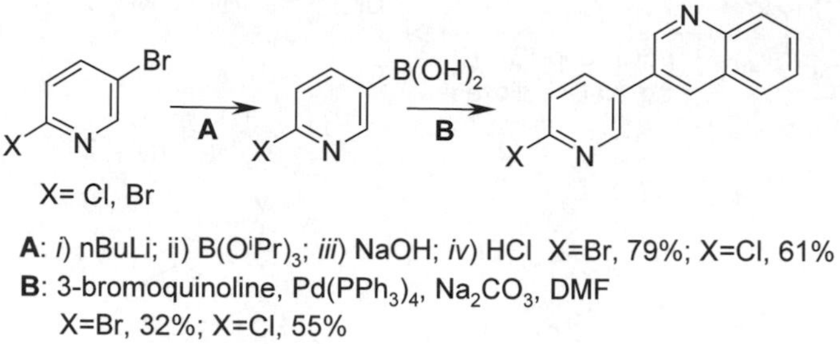
Br
B(OH)$_2$
A
B
X
N
X= Cl, Br
A: *i*) nBuLi; ii) B(O^{i}Pr)$_3$; *iii*) NaOH; *iv*) HCl X=Br, 79%; X=Cl, 61%
B: 3-bromoquinoline, Pd(PPh$_3$)$_4$, Na$_2$CO$_3$, DMF
X=Br, 32%; X=Cl, 55%

(7.5.)

In an analogous case (**7.6.**) the 9-borabicyclo[3.3.1]nonane (BBN) unit was introduced onto the quinoline core through lithium-halogen exchange and the quenching of the intermediate with 9-BBN-OMe. The quinolylborane coupled with 3-bromopyridine under the earlier mentioned conditions (THF, KOH, TBAB) to give the pyridylquinoline in acceptable yield.[8]

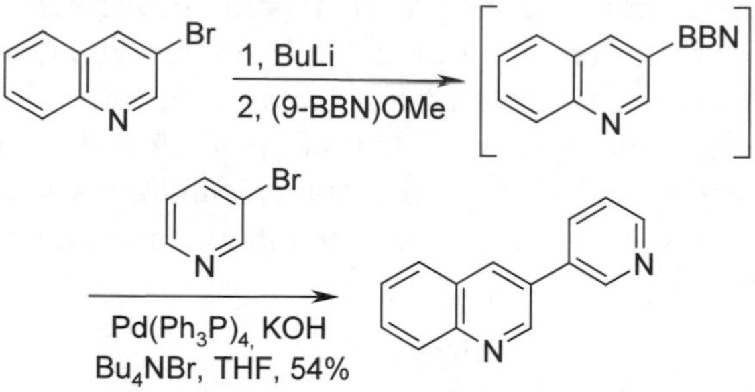
Br
1, BuLi
2, (9-BBN)OMe
BBN
N
Pd(Ph$_3$P)$_4$, KOH
Bu$_4$NBr, THF, 54%

(7.6.)

An alternate approach to the formation of pyridylboronic acids is the cross-coupling of a halopyridine with a diboronate ester (usually bis(pinacolato)diboron, **7.7.**).[9] The analogous reaction of 2-chloropyridine led to pyridine formation through protodeboronation. The product of the reaction, either after hydrolysis to the boronic acid or in the ester form, can be further reacted with another aryl halide to give a biaryl. In certain cases the reaction might also be carried out in a one-pot manner.[10]

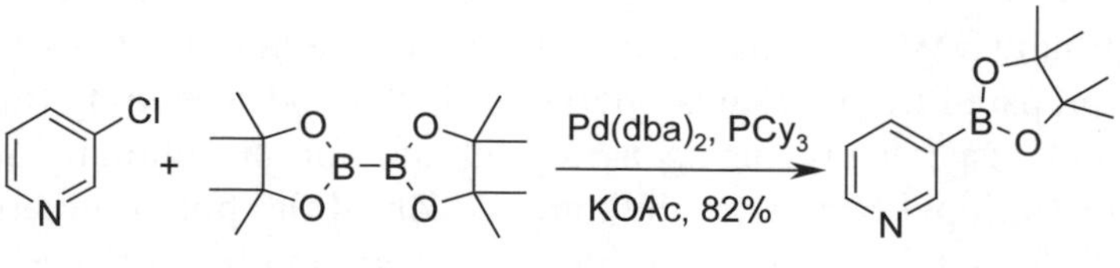
Cl
+
B–B
O
N
Pd(dba)$_2$, PCy$_3$
KOAc, 82%

(7.7.)

In the presence of more than one halogen in the heterocyclic core the issue of selectivity arises. In general halogen atoms at activated positions (*e.g. ortho* or *para* to ring nitrogen atoms) are favoured in the coupling as demonstrated by Timári and co-workers in the synthesis of Quindoline.[11] If the two halogen atoms are in equivalent position, then the direction of the coupling is determined by steric factors or secondary interactions. The regioselectivity in the Suzuki coupling of 2,6-dichloronicotinamides with arylboronic acids (**7.8.**) in the presence of Li's Pd/phosphinous acid catalyst,[12] for example, was governed by the chelation of the boronic acid to the amide group and the 2-arylated pyridine derivatives were isolated in good yield.[13]

$$R=NHCH_2CH_2OPh$$

(7.8.)

Dihaloazines might also undergo double coupling in the presence of excess boronic acid. 2,6-Dichloropyrazines were reported to give 2,6-diarylpyrazines in excellent yield even with such sterically demanding partners as 2,6-dimethyl-phenylboronic acid (**7.9.**). Interestingly, when 2,3–dichloropyrazine was reacted with the same boronic acid only the formation of the monoarylated pyridazine was observed. The absence of the second coupling was attributed to the steric bulk of the introduced aryl group.[14]

(7.9.)

Diazines (pyridazines, pyrimidines and pyrazines) undergo cross-coupling more readily than pyridines, due to the electron deficient nature of the heterocyclic ring. 4-Bromopyridazines bearing electron donating substituents in position 3 were found to couple readily with arylboronic acids in the presence of tetrakis(triphenylphosphane) palladium (**7.10.**).[15]

i) HBr, AcOH, sealed tube
ii) ArB(OH)$_2$, Pd(PPh$_3$)$_4$, Na$_2$CO$_3$, EtOH-PhMe

(7.10.)

The Suzuki-coupling of a 4-iodopyridazinon derivative with an aniline derived boronic acid was exploited in the preparation of the indolopyridazine skeleton (**7.11.**).[16]

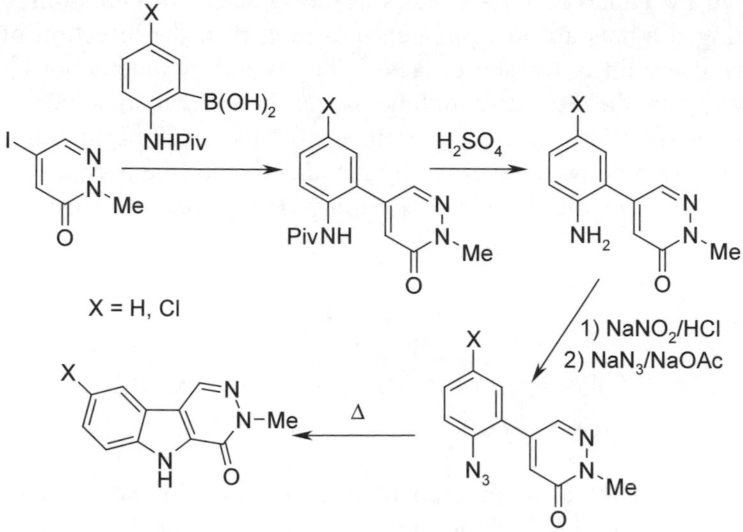

(7.11.)

The synthesis of diazineboronic acids, for example pyrimidine 5-boronic acid (**7.12.**), is more challenging. Gronowitz succeeded in its preparation, by maintaining very low temperatures during the lithium-bromine exchange reaction to minimize competing reactions.[17] The boronic acid was coupled with tropolone derivatives in modest yield in the search for inositol monophosphatase inhibitors.[18]

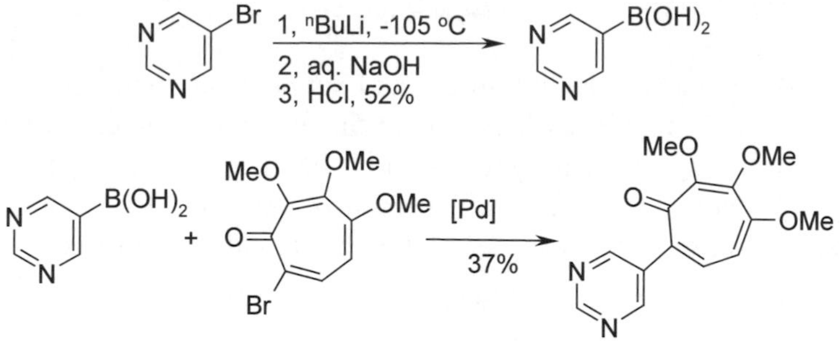

(7.12.)

The introduction of a uracil moiety onto thiophene was achieved by the use of a similar boronic acid, bearing *tert*-butoxyde substituents in positions 2 and 4 (**7.13.**). This compound was reacted with a bromothiophene to give the Suzuki coupling product in 63% yield, which on treatment with hydrochloric acid released the *tert*-butyl groups to furnish the desired thienyl-uracil.[19]

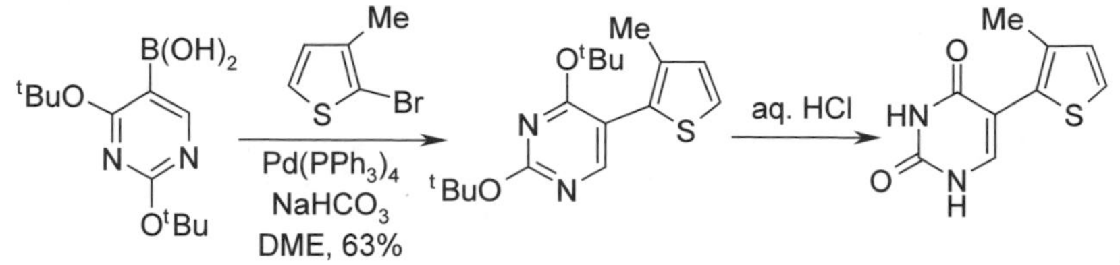

(7.13.)

In the coupling of 2-metylthio-5-bromo-3H-pyrimidine-4-on derivatives selective reaction was achieved in position 2 or 5 (**7.14.**) through the selection of the reagents. Under regular coupling conditions the oxidative addition occurred on the carbon-bromine bond and the organic moiety of the boronic acids was introduced into position 5 of the pyrimidine ring. The same compound, in the presence of 1.5 equivalent of a copper(I) salt underwent oxidative addition on the carbon-sulphur bond and the Suzuki coupling gave 2-arylpyrimidones exclusively. The effect of the copper additive was attributed to its marked affinity towards sulphur. The selection between bromine and sulphur allowed for the selective introduction of different boronic acids onto the pyrimidine core.[20]

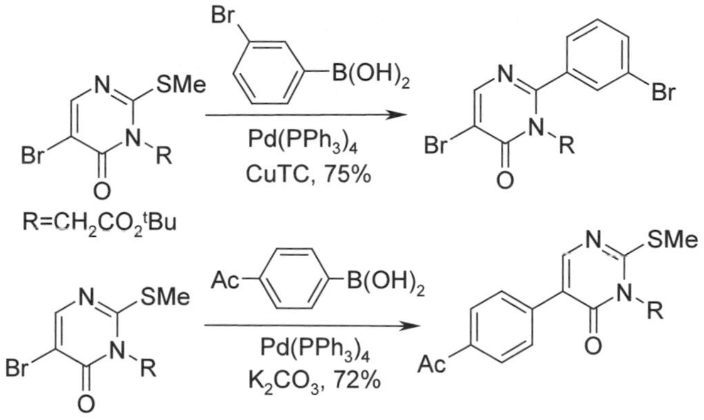

(7.14.)

Nguyen demonstrated the efficiency of the Suzuki coupling in a comparative study.[21] The Suzuki coupling of pyridine 4-boronic acid with a 4-bromo-salicyl alcohol derivative gave the desired product in 74% yield (**7.15.**). Reaction of the same bromo compound and 4-tributylstannyl-pyridine (Stille coupling) furnished the same product in only 28%.

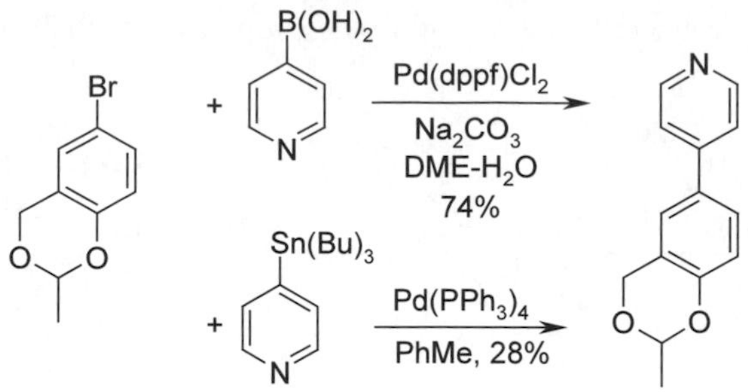

(7.15.)

Kharasch (Kumada) coupling

The transition metal catalyzed carbon-carbon bond formation between organomagnesium reagents and aryl (vinyl) halides has been one of the pioneering entries into cross-coupling chemistry. The reaction has been widely utilized since than in azine chemistry,[22] with the limitation that the functional group tolerance of Grignard reagents is only moderate. Here only some of the more recent developments will be mentioned.

The Kharasch coupling, unlike most of the other cross-coupling reactions, is usually run in the presence of a nickel catalyst. The effect of the ligands on the efficiency of the coupling was studied recently in the reaction of halopyridines with Grignard reagents using a nickel based catalyst, and nucleophilic carbenes derived from imidazolium salts were found to be the most effective.[23] Under the optimised conditions 2-chloropyridine coupled readily with 4-methoxyphenylmagnesium bromide already at ambient temperature to yield 2-anisylpyridine (**7.16.**)

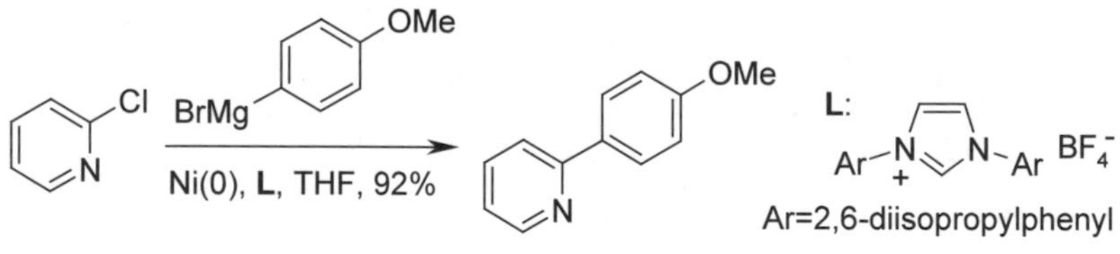

(7.16.)

The ease of coupling, originating in the combination of nickel(0)'s willingness to undergo oxidative addition and the enhanced transmetalation ability of Grignard reagents, led to the first cross-coupling reactions of fluorinated heterocycles.[24] Fluoroazines and fluorodiazines reacted with arylmagnesium bromides (**7.17.**) in the presence of a bidentate phosphine based nickel catalyst to give the coupled products in good to excellent yield.

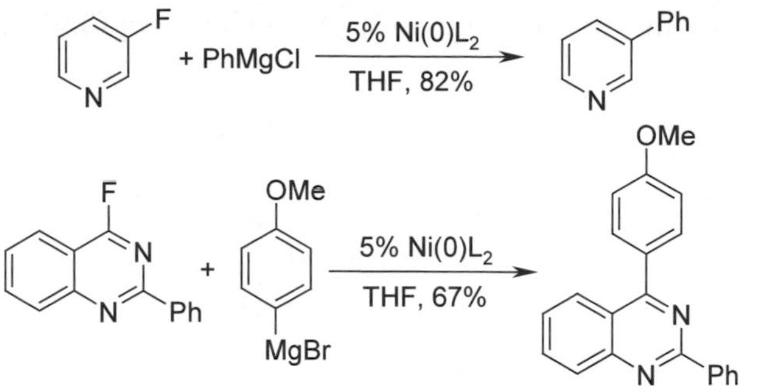

(7.17.)

A new variant of the cross-coupling of Grignard reagents and heteroaryl halides utilizing cobalt as a catalyst was reported recently.[25] The reaction, which is limited to date to 2-chloropyridine and related heterocycles, proceeds readily at low temperatures (sometimes at -40 °C), which allows for the use of functionalized Grignard reagents too. The coupling of 2-chloroquinoline and 1-chloroisoquinoline with phenylmagnesium bromide was found to run smoothly in the presence of 5 mol% cobalt or iron powder too (**7.18.**).

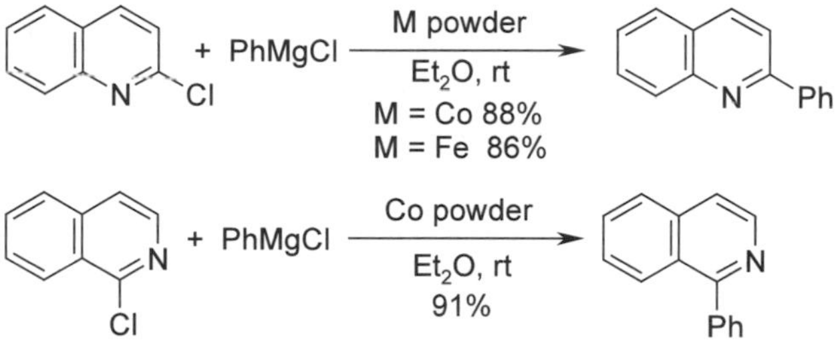

(7.18.)

Negishi coupling

Organozinc reagents represent a midway in cross-coupling chemistry between the less functional group tolerant but readily transmetalating Grignard reagents and organoboron reagents that show high functional group tolerance but transmetalate only moderately. The preparation of organozinc reagents is achieved either by direct zinc insertion into a carbon-halogen bond, or through the transmetalation of the appropriate organometallic reagent with an anhydrous zinc salt. This later approach is frequently coupled with directed *ortho*-lithiation providing a convenient entry to functionalised biaryls. In a representative example (**7.19.**) benzoic acid diisopropylamide was converted into the appropriate organozinc reagent in a lithiation-transmetalation sequence, and was coupled with pyridine-3-triflate in the presence of a nickel catalyst to give the desired 3-pyridyl derivative in 89% yield.[26]

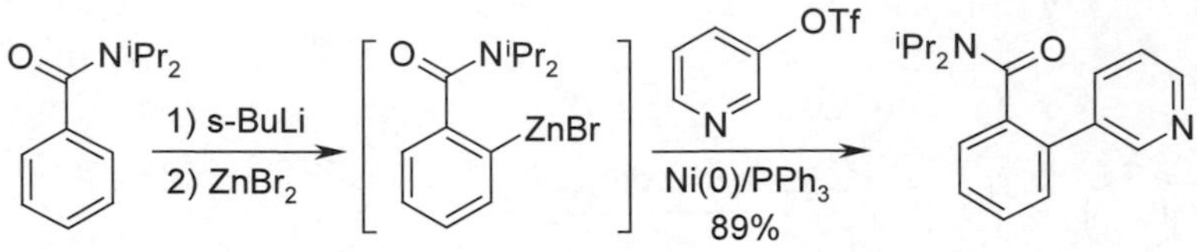

(7.19.)

2-Pyridylzinc bromide, a commercially available reagent, was used in the synthesis of a series of bipyridyl derivatives (**7.20.**). Activated coupling partners, such as 2-chloro-4-cyanopyridine, reacted readily at room temperature in the presence of a palladium based catalyst, while less electron deficient halopyridines were coupled at elevated temperatures.[27] The same approach was also extended to the preparation of other heterobiaryls.[28]

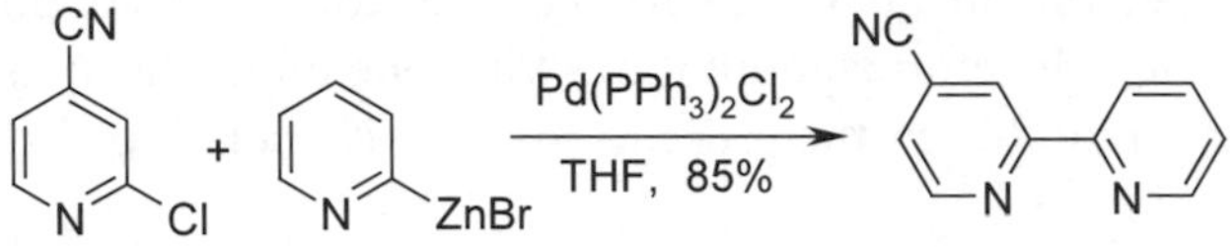

(7.20.)

The effect of the catalyst source and the mode of heating on the coupling of a pyridylzinc halide, prepared from 2-fluoro-4-iodopyridine through lithium-iodine exchange and transmetallation, and 2,4-dichloropyrimidine was studied by Stanetty.[29] The use of tetrakis(triphenylphosphino)palladium led to the formation of the 4-pyridylpyrimidine derivative with good selectivity (**7.21.**), while microwave irradiation allowed for the selective coupling at one or both positions leading to the 4-pyridylpyrimidine or 2,4-dipyridylpyrimidine respectively in a relatively short time. The change of the catalyst to Pd/C (no ligand added) led to the predominant dimerization of the pyridylzinc reagent.

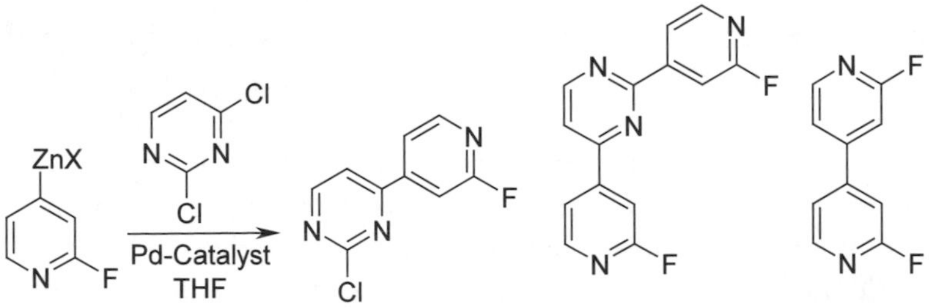

Catalyst	A	B	C
Pd(PPh$_3$)$_4$, Δ	90%	5%	0%
Pd(PPh$_3$)$_4$, MW	90%	0%	4%
Pd/C, MW	18%	70%	0%

(7.21.)

Quéguiner used two subsequent Negishi-couplings on a pyridine ring in the total synthesis of the antibiotics Caerulomycin B and C. The appropriate 3,4-dihydroxypyridine derivative was converted to a pyridylzinc derivative through lithiation-transmetalation (**7.22.**), and coupled with 2-bromopyridine. The resulting trisubstituted pyridine was halogenated and the bromine was exchanged to a methyl substituent in another Negishi coupling with methylzinc chloride providing a key intermediate for both Caerulomycins.[30]

(7.22.)

Stille coupling

Prior to the recent emergence of Suzuki reaction as the method of choice for the coupling of substrates with sensitive functional groups, the Stille coupling was probably the predominant and most versatile transition metal catalyzed transformation. The functional group tolerance of organostannanes and their stability allows the isolation of such intermediates. The common problems associated with Stille coupling are the increased toxicity of the used organotin reagents and the difficulty encountered during the removal of the tin containing side products from reaction mixtures.

The coupling of haloazines and arylstannanes usually proceeds readily, analogous to other Stille couplings. This compilation contains only one recent example, where the outstanding activity of a palladium based catalyst system was extended to the Stille coupling of 4-chloroquinoline derivatives with trimethylphenyltin (**7.23.**). Analogous to the Suzuki coupling, the use of some additive enhancing the transmetalating ability of the organometallic reagent is beneficial. In the reported examples the addition of caesium fluoride[31] led to an increase in the yield of the product.[32]

(7.23.)

The stability of hetaryltin compounds makes them useful reagents in cross-coupling reactions. Unlike with pyridylboronic acids, the position of the stannyl group on the ring has only a minor influence on its reactivity. The most common way for the preparation of trialkylstannyl-azines is through the reaction of the lithiated heterocycle with the appropriate trialkyltin chloride.[33] An alternate approach utilizes the ability of hexaalkylditins (*i.e.* hexabutylditin) to transfer a trialkylstannyl group onto an aromatic core in a formal Stille reaction with an aryl halide. The resulting aryltrialkyltin compound can either be isolated[34] or used in a subsequent coupling. Heating a 2:1 mixture of 2,5-dibromopyridine and hexabutylditin in the presence of a palladium catalyst, for example, led to the isolation of the symmetrical bipyridine derivative in 80% yield (**7.24.**).[35] The observed selectivity of the coupling is attributed to the enhanced reactivity of the 2-position compared with position 5.

$$\text{(7.24.)}$$

The use of bifunctional pyridine derivatives allows for the introduction of two aryl moieties onto the heterocyclic core in the same reaction. The preparation of chelating 2,2':6',2"-terpyridines (**7.25.**) was achieved both in the Stille coupling of 2,6-dibromopyridine with 2-tributylstannylpyridines and in the coupling of 2,6-bis(trimethylstannyl)pyridine with 2-bromopyridines using the same tetrakis(triphenylphosphine)palladium catalyst. The two routes gave consistent results in most cases (43-90% yield depending on the substitution pattern) with the exception of the coupling of 2,6-dibromopyridine with 6-methyl-2-tributylstannylpyridine which stopped at the bipyridine form.[36]

$$\text{(7.25.)}$$

The introduction of a masked uracil moiety onto the pyridine core was achieved in the reaction of 3-tributylstannylpyridine with 5-bromo-2,4-bis(trimethylsilyloxy)pyrimidine (**7.26.**) The coupling proceeded in the presence of bis(triphenylphosphine)palladium dichloride to give the desired product in 42% yield (*c.f.* **7.13.**).[37]

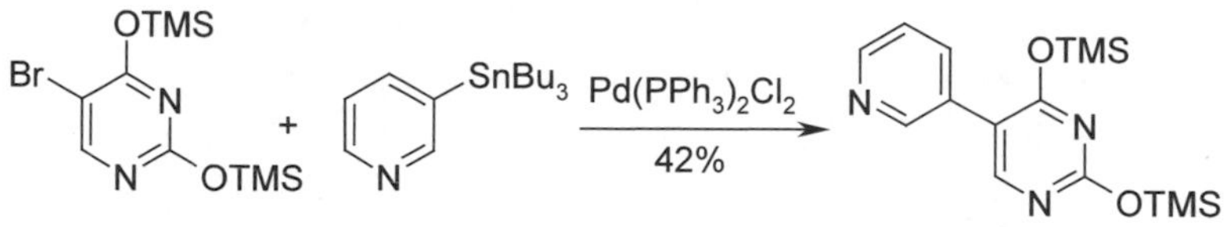

(7.26.)

The Stille coupling of pyridylstannanes was also extended to stannylpyridinium salts. 1-Methyl-3-tributylstannylpyridinium tosylate was coupled with 2-chloropyrazine in excellent yield (7.27.). Using the corresponding *N*-oxide the same coupling gave only 29% yield.[38]

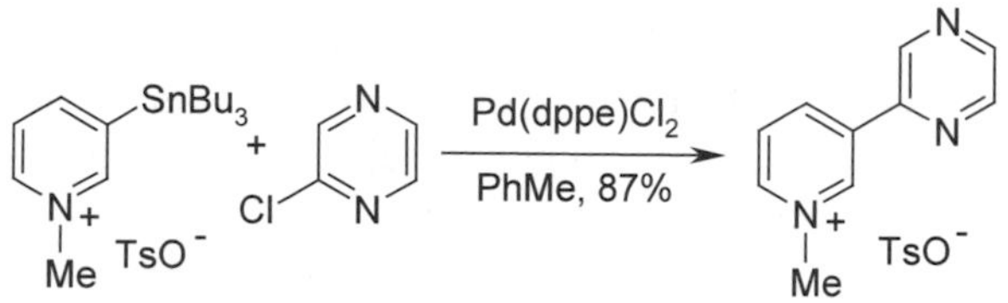

(7.27.)

The functionalization of the chroman skeleton was achieved by the Stille coupling of a chromenyl triflate, derived from the appropriate chromanon derivative, and 2-trimethylstannylpyridine (7.28.). In the coupling of the chromenyl triflate the addition of one equivalent of lithium chloride to the reaction mixture was found to facilitate the process, probably making the intermediate chromenylpalladium triflate complex more willing to transmetalate through ligand exchange. The 2'-pyridyl-*N*-oxide moiety was also introduced onto the same heterocyclic core using 2-pyridylzinc chloride *N*-oxide in a Negishi coupling protocol in 64% yield.[39]

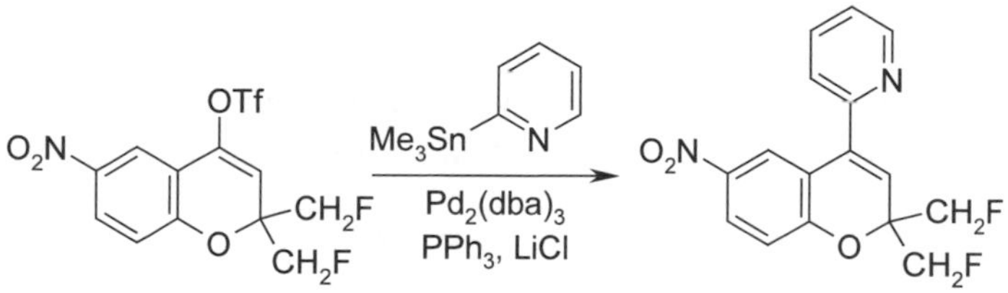

(7.28.)

The Stille coupling of 2-chloro-5-tributylstannylpyridine with an enantiopure 2-iodo-cyclohexenon (7.29.) derivative formed the basis of the total synthesis of (+)-epibatidine. The reaction is a nice example of the chemical inertness of arylstannanes and the mildness of the coupling conditions. Both the enone moiety and the chiral allylic centre remained untouched in the process. The effective coupling required the use of a soft ligand, triphenylarsine and the addition of copper(I) iodide as co-catalyst.[40]

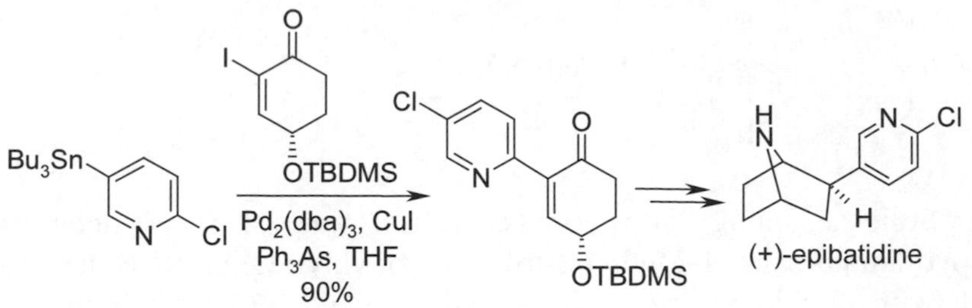

(7.29.)

Another example of the mild nature of the Stille coupling and the functional group tolerance of hetarylstannanes is the reaction of 5-tributylstannylpyrimidine with diphenyl 4-bromo-2-nitrobenzylphosphonate (**7.30.**). The coupling reaction, conducted in refluxing acetonitrile gave the biaryl derivative in 65% yield.[41]

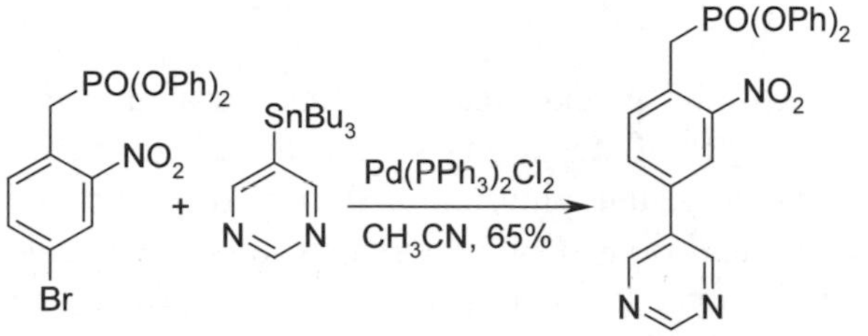

(7.30.)

An "exotic" example demonstrating the scope of Stille coupling is the transformation of 2,4,6-tribromo-3-methylphosphinine with 2-tributylstannylpyridine and 2-trimethylstannylfurane (**7.31.**). The short lived coupling products were isolated in 40% and 60% yield respectively, the furylstannane resulting in the formation of the 2,6-disubstituted product. The palladium catalyzed coupling of (trimethylsilyl)diphenylphosphine with the same tribromophosphinine also gave 2,6-disubstitution.[42]

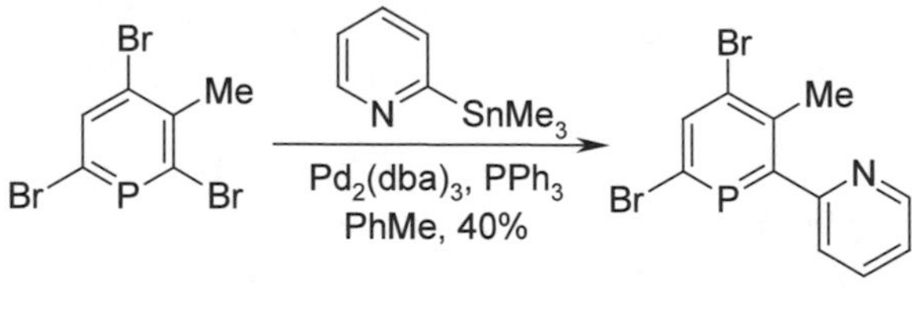

(7.31.)

Sonogashira coupling

The reaction of haloazines and terminal acetylenes in the presence of a palladium(0)-copper(I) catalyst system usually proceeds readily. The only problem encountered in such reactions is the undesired dimerization of the acetylene, which might become dominant in slow processes. Unlike in the case of aromatic carbocycles, where the coupling is in most cases limited to aryl iodides and bromides, chloroazines are usually coupled with success too.

The coupling of 3-bromopyridine and phenyl propargyl alcohol was already achieved at room temperature if triethylamine was used as base and solvent. Interestingly, under the same conditions a 3-chloro-5-trifluoromethyl-2-iodopyridine gave an isomeric product (**7.32.**). The rearrangement, whose product is analogous to the result of a Heck coupling, is uncommon in Sonogashira coupling.[43]

The Sonogashira coupling of halopyridines was also achieved by other means. Resin bound bromopyridine derivatives, for example, underwent smooth coupling with acetylenes as well as with arylboronic acids and aryltin reagents.[44] The advantageous effect of microwave irradiation on the coupling of halopyridines and trimethylsilylacetylene was also reported.[45]

(7.32.)

By using bifunctional azines one might be able to introduce two acetylene moieties sequentially or simultaneously. The coupling of 3-trifluoromethylsulfonyloxy-2-bromopyridine with trimethylsilylacetylene, for example, gave the diethynylated pyridine compound in reasonable yield (**7.33.**). The use of this acetylene derivative allows for the preparation of arylacetylenes too, as the TMS group is readily removed on treatment with fluoride ions. This way 2,3-diethynylpyridine was prepared in a coupling-deprotection sequence.[46] 2,6-Diethynylpyridine was prepared in an analogous process starting from 2,6-dibromopyridine.[47]

An alternate reagent that was used for the introduction of a masked acetylene moiety onto the pyridine core is 2-methyl-3-butyne-2-ol. Under appropriate conditions the first Sonogashira coupling and the removal of the protecting group can be realized in the same pot, allowing for the subsequent introduction of another aryl group onto the freed acetylene carbon. This domino coupling procedure was used in the one-pot transformation of 3-bromopyridine into the disubstituted acetylene derivative shown in **7.34.**[48]

(7.33.)

(7.34.)

The Sonogashira coupling of haloazines can be effected by a series of catalyst systems. Recently a lot effort was devoted to the development of a recyclable catalyst system. Kotschy and co-workers recently reported the use of palladium on charcoal as a convenient palladium source for this process, which allows for the separation and reuse of the catalyst at the end of the reaction (**7.35.**). The authors also demonstrated that, in spite of the absence of any substantial catalyst leaching, the catalytic activity of the reused Pd/C decreases on each run,[49] a surprising phenomenon which was attributed to the dissolution and reprecipitation of the active catalyst in the course of the process. Pd(OH)$_2$ on charcoal exhibited a similar activity in the Sonogashira coupling of bromopyridines.[50]

ArX: 2-bromopyridine; **a**: 58%; **b**: 62%; **c**: 65%   **a**: R = C$_2$H$_4$OH
ArX: 3-bromopyridine; **a**: 77%; **b**: 81%; **c**: 69%   **b**: R = C$_4$H$_9$
ArX: 2-chloropyridine; **b**: 54%; **c**: 51%   **c**: R = C(CH$_3$)$_2$OH

(7.35.)

Like halopyridines, diazines participate in Sonogashira coupling too. 3,6-dimethyl-2-chloropyrazin, for example on coupling with phenylacetylene under standard conditions, gave the desired compound in good yield, which was further reduced to give a natural product (**7.36.**)[51] (*N.B.* the Heck reaction, which could be considered as an alternate approach would be expected to furnish predominantly the *cis*-olefin as product).

A systematic survey of the Sonogashira coupling of 2-halopyrimidines under mild conditions revealed that, in line with the expectations, the iodopyrimidines give excellent yield, while the reactivity of bromopyrimidines is mediocre and chloropyrimidines hardly react (**7.37.**).[52] These results have, of course only limited relevance as increase of the reaction temperature might drive the coupling of the latter halopyrimidines to completion too. An example of the increased reactivity of chlorodiazines is presented in **7.38.** where, *en route* to a PDE4 inhibitor a chlorophthalazine derivative was coupled efficiently with phenylacetylene.[53]

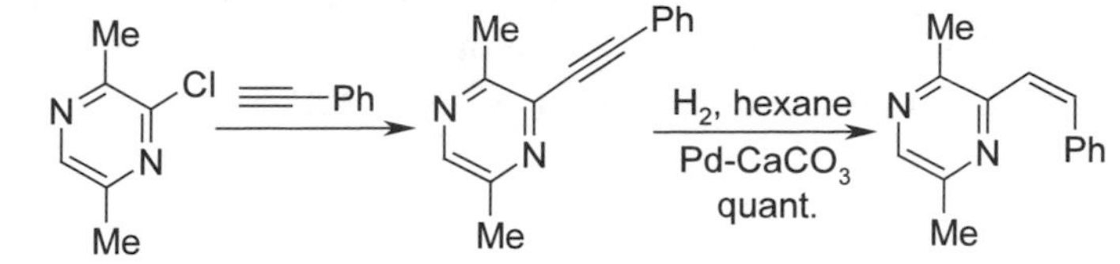

(7.36.)

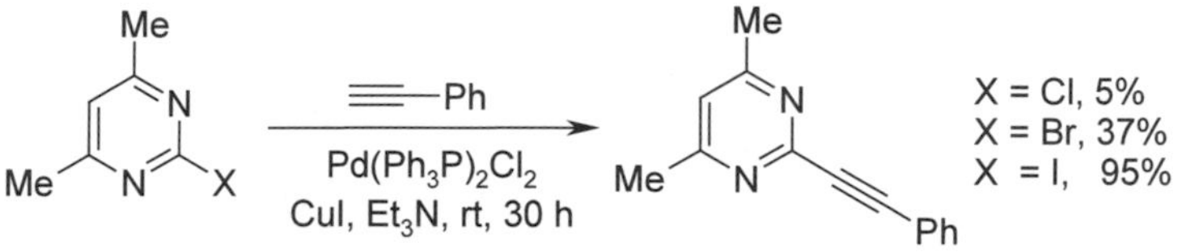

X = Cl, 5%
X = Br, 37%
X = I, 95%

(7.37.)

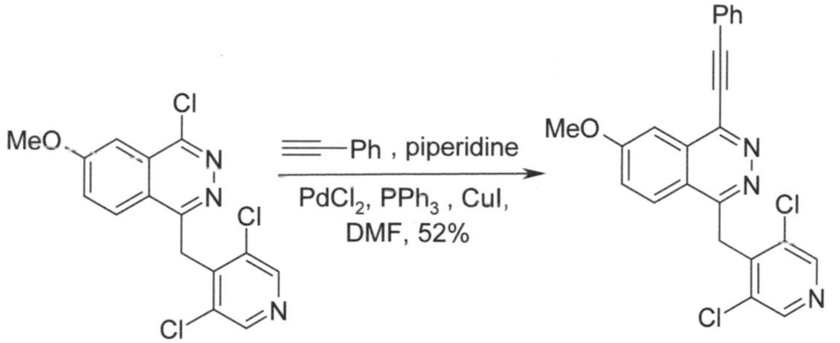

(7.38.)

The Sonogashira coupling was reported to succeed in difficult cases too, where the other coupling reactions failed so far. Chlorotetrazines were coupled successfully for the first time with different acetylene derivatives in acceptable yield (**7.39.**).[54] The same chlorotetrazines were too sensitive to undergo Suzuki, Karasch or Negishi coupling with aryl groups, and in those cases only side reactions were observed[55] (*N.B.* the analogous triazine derivatives were shown to participate both in Sonogashira[56] and Negishi[57] couplings).

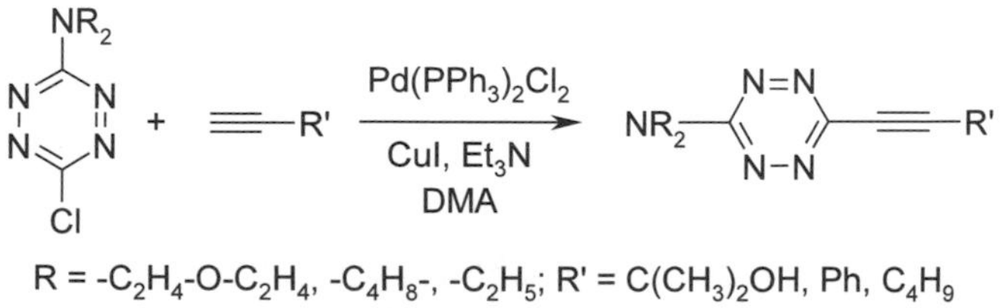

R = -C$_2$H$_4$-O-C$_2$H$_4$-, -C$_4$H$_8$-, -C$_2$H$_5$; R' = C(CH$_3$)$_2$OH, Ph, C$_4$H$_9$

(7.39.)

The chlorotetrazines were found to react not only under regular Sonogashira coupling conditions, but also in the presence of zinc acetylides. This reaction is an extension of the Negishi coupling, which sometimes gives superior results to the Sonogashira coupling. In a representative application 1,8-diethynylanthracene was coupled with a 4-iodopyridine derivative in the presence of a palladium catalyst, replacing the copper(I) iodide co-catalyst with a stoichiometric amount zinc(II) triflate to avoid undesired dimerisation of the acetylene derivative (**7.40.**).[58]

(7.40.)

Another alternate to the Sonogashira coupling was reported by Blum and Molander, where sodium tetraalkynylaluminates were coupled with bromoazines and bromoazoles in the presence of a palladium-triphenylphosphine catalyst system. 5-Bromopyrimidine coupled with the TMS-acetylide, for example, to give the ethynylpyrimidine in excellent yield (**7.41.**).[59] The transmetalating reagents were prepared in situ by the reaction of the appropriate acetylene derivative with sodium aluminiumhydride.

(7.41.)

The cross-coupling of 3-iodopyridine and 4-dimethylamino-phenylacetylene was reported to work efficiently in the presence of a nickel based catalyst system too (**7.42.**).[60] The described conditions (*e.g.* catalyst loading, solvent, temperature, additive) are more or less the same as in the conventional palladium catalyzed variant, although the nickel based system gave only poor results with bromoazines.

(7.42.)

An analogous nickel catalyzed coupling was also reported recently, where 3-, and 4-cyanopyridine were coupled with different ethynylzinc derivatives in the presence of a nickel-phosphine catalyst system (**7.43.**).[61] Although this reaction is not a Sonogashira coupling, it constitutes an efficient alternative approach to ethynylpyridines. It is also interesting to note, that in this coupling the cyano group serves as a halide surrogate and in the first step of the catalytic cycle nickel inserts into a carbon-carbon bond.

$$\text{Pyridine-CN} + \text{Ph}\!-\!\!\!\equiv\!\!\!-\text{ZnBr} \xrightarrow[\text{THF, 45\%}]{\text{Ni(PMe}_3)_2\text{Cl}_2} \text{Pyridine}\!-\!\!\!\equiv\!\!\!-\text{Ph}$$

(7.43.)

Hiyama coupling

The palladium catalyzed cross-coupling of organosilicon compounds and aryl halides found only limited application with azines compared to the Suzuki or Negishi coupling. In a recent paper DeShong reported the efficient coupling of bromopyridine derivatives with aryl siloxanes (**7.44.**).[62] The transmetalating ability of the siloxane was enhanced by the addition of tetrabutylammonium fluoride.

$$\text{(MeO, Me-pyridine-Me, Br)} + \text{Ph(SiOMe)}_3 \xrightarrow[\substack{\text{TBAF, DMF} \\ \text{97\%}}]{\text{Pd(OAc)}_2, \text{PPh}_3} \text{(MeO, Me-pyridine-Me, Ph)}$$

(7.44.)

7.2   INSERTION ROUTE

The halogenated derivatives of six membered heterocycles, like their carbacyclic analogues, usually participate readily in coupling reactions that involve the incorporation of an olefin or carbon monoxide. The insertion of carbon monoxide commonly leads to the formation of either a carboxylic acid derivative or a ketone, depending on the nature of the other reactants present. Intermolecular and intramolecular variants of the insertion route are equally popular, and are frequently utilized in the functionalization of heterocycles or the formation of annelated ring systems.

Heck reaction (olefin insertion)

The palladium catalyzed coupling of haloazines and olefins is a robust process, which is usually run at an elevated temperatures in the presence of a "simple" catalyst and at least an equimolar amount of base to neutralise the formed hydrogen halide. The presence of a ring heteroatom might in certain cases lead to complex formation, which deactivates the catalyst and kills the process.[63] The regioselectivity of the coupling is predominantly governed by steric and electronic factors. While on electron deficient olefins such as acrylates the aryl group is preferentially attached to the β-carbon, in the case of the more electron rich styrenes and vinyl ethers the formation of product mixtures is sometimes observed due to the competition of steric and electronic control.

The *N*-protected 3-amino-2-chloropyridine derivative in **7.45.** for example was found to react with styrene in the presence of a palladium acetate-triphenylphosphine catalyst system at elevated temperatures to give a near quantitative yield of a single coupling product.[64] In the process sodium acetate was used as base. The formation of a single regioisomer might be attributed to the steric bulk of pyridine's substituents in position 3.

$$(7.45.)$$

In the coupling of 3-(2'-chloropyridyl)-triflate and ethyl acrylate (**7.46.**) steric and electronic factors both drive the pyridine moiety into the β-position of the olefin.[65] Although the 2-position of the pyridine ring is more activated, the enhanced reactivity of the triflate in oxidative addition leads to selective reaction in the 3-position. Analogous to some previously mentioned reactions involving triflates the addition of lithium chloride was found to accelerate the coupling (*c.f.* **7.28.**).

$$(7.46.)$$

Strained olefins usually undergo insertion more readily. In certain cases, however, the product of the carbometalation step is unable to undergo the concluding β-hydride elimination due to geometric constrains. One such example is the coupling of 2-chloro-5-iodopyridine with an azabicycloheptenone derivative (**7.47.**) giving rise to two regioisomers.[66] Under the applied conditions sodium formate serves the purpose of converting the intermediate formed in the carbopalladation to the desired product through a ligand exchange–CO_2 extrusion–reductive elimination sequence. It is also interesting to note that in the process tetrabutylammonium chloride replaces common phosphine ligands to stabilize the catalytically active palladium species in a colloidal form.

Ethylene might also be used in Heck coupling as was demonstrated in the kilogram scale preparation of a vinylpyridine derivative. *N*-Acetyl 2-amino-

5-bromopyridine was coupled with ethylene in the presence of a catalyst consisting of palladium acetate, tri-*o*-tolylphosphine and BINAP, and triethylamine (**7.48.**).[67]

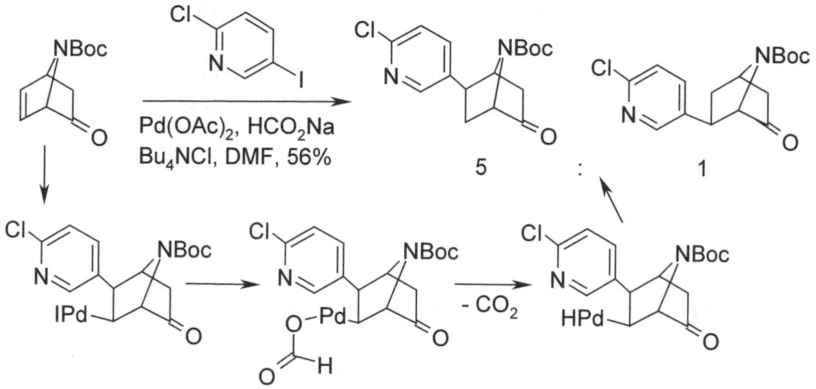

(7.47.)

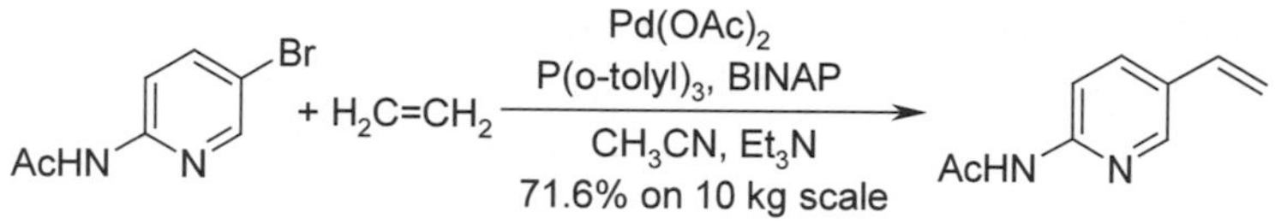

(7.48.)

The proximity of an aryl substituent to the halogen atom on the heterocycle might lead to undesired side reactions in certain cases. The Heck coupling of 4-(*p*-nitrophenyl)-3-bromopyridine with ethyl acrylate for example gave a 3:1 mixture of the expected pyridylacrylate and another product where dehalogenation of the pyridine moiety was coupled with the introduction of the acrylate onto the phenyl ring (**7.49.**).[68] Formation of the unexpected product was rationalised by assuming a hydrogen-palladium exchange reaction between the neighbouring aromatic rings (through a palladacyclic intermediate). Interestingly, the coupling of the analogous 3-(*p*-nitrophenyl)-4-brompyridine and ethyl acrylate gave only the expected product (**7.49.**) and no rearrangement was observed.[69]

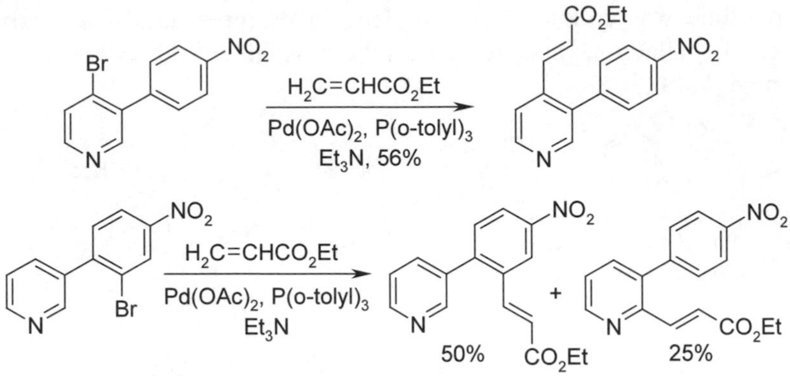

(7.49.)

The Heck reactions depicted so far all involve the coupling of halopyridines and other olefins. The alternate approach, coupling of a vinylpyridine with an aryl halide is also feasible, although less commonly employed. 4-Vinylpyridine was coupled successfully with diethyl 4-bromobenzylphosphonate (7.50.) in the presence of a highly active catalyst system consisting of palladium acetate and tri-*o*-tolylphosphine to give the desired product in 89% yield, which was used for grafting the pyridine moiety onto metal oxides.[70]

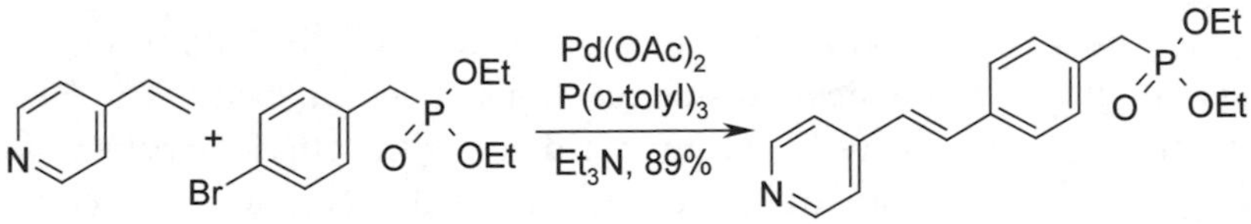

(7.50.)

If the olefin moiety and the haloazine are covalently attached, then the Heck reaction leads to the formation of a condensed ring system. This strategy is commonly employed in natural product synthesis as shown on the example of (±)-oxerine (7.51.). Here the annelating ring was established through the intramolecular coupling of a 4-homoallyl-3-bromopyridine derivative. In the reaction the exclusive formation of the five membered ring was observed.[71]

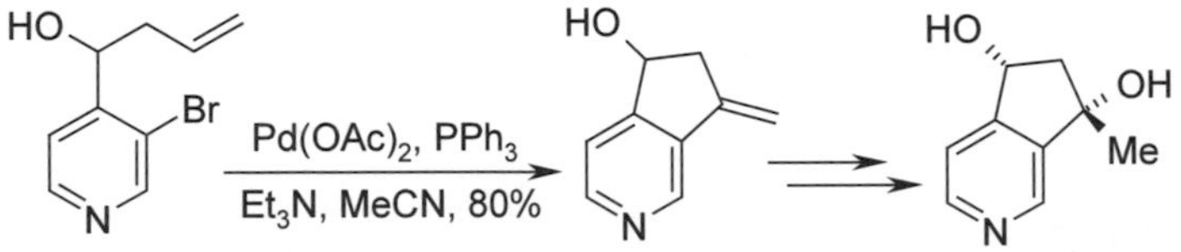

(7.51.)

A similar strategy was followed in the synthesis of bridged pyridotropane derivatives (7.52.). The iodopyridine and acrylate subunits were both born by a pirrolidine ring. The *cis*-alignment of the two substituents is crucial for the success of the coupling, which led to the selective formation of the seven membered cyclic product.[72]

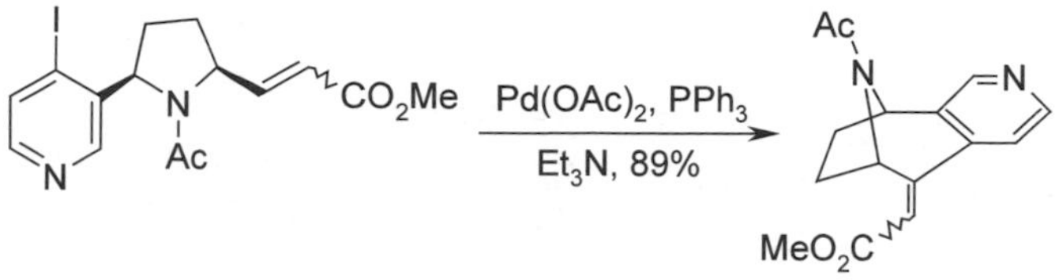

(7.52.)

Halogenated diazines are usually more reactive in coupling reactions than the corresponding pyridine derivative. 2,6-Dimethyl-4-iodopyrimidine, for example, was found to react with ethyl acrylate in triethylamine in the presence of palladium on charcoal to give the β-pyrimidylacrylate in good yield (**7.53.**).[73] Although this reaction is formally catalyzed by heterogeneous palladium, recent studies suggest, that the active catalyst is homogeneous: palladium is dissolved at the beginning of the coupling, and is re-deposited onto the charcoal at the end of the process.[74]

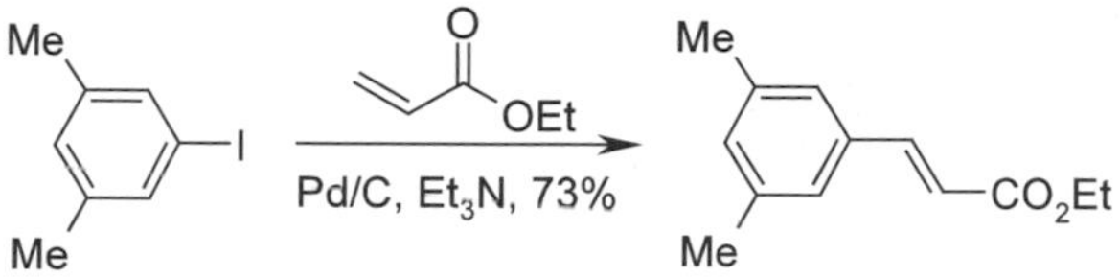

(7.53.)

The Heck coupling of haloazines and cyclic unsaturated sugar derivatives provides a convenient access to C-nucleosides. A silylated ribofuranosid glycal was coupled with a substituted iodopyrazine (**7.54.**) in the presence of a palladium-triphenylarsine catalyst system to give the desired product in good yield, which was converted into the nucleoside in a deprotection-reduction sequence. The reactivity of the dihydrofuran skeleton makes this approach an attractive route for the introduction of the biologically important deoxyribofuranosyl moiety.[75]

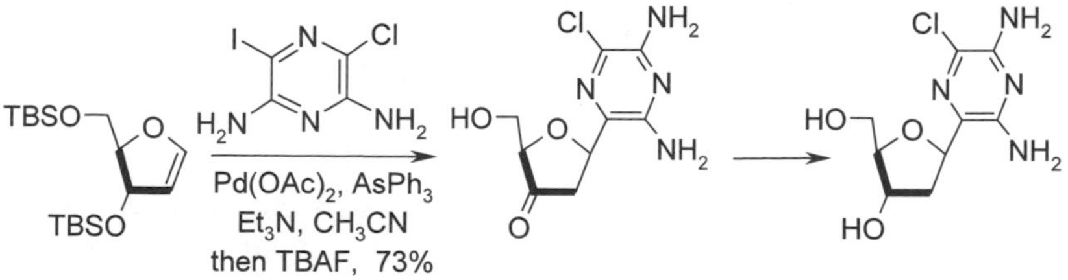

(7.54.)

Compounds containing cumulated double bonds (e.g. allenes) also undergo Heck coupling with haloazines. 5-Bromopyrimidine was found to react with an acetylallene to give rise to the furylated pyrimidine in near quantitative yield (**7.55.**)[76]. Here the carbopalladation takes place on the double bond further away from the acetyl moiety, followed by the formation of a carbon-oxygen bond giving rise to the 3-hetarylfuran derivative.

(7.55.)

The furan moiety can be introduced onto an azine in a different manner too. Aryl halides, including haloazines, can undergo a formal Heck reaction with furan, also known as the heteroaryl Heck reaction, to give the 2-hetaryl-furan derivative. 2-Chloro-3,6-dimethylpyrazine, for example, reacted with furan to give 2-(3',5'-dimethylpyrazyl)-furan in good yield (**7.56.**).[77] *N.B.* the heteroaryl Heck reaction is complimentary to the approach outlined in **7.55.** as it gives the regioisomeric product selectively.

(7.56.)

The heteroaryl Heck reaction is an efficient tool for the introduction of other five membered heteroaromatic systems too (for more details see Chapter 6.4.). Chloropyrazines reacted readily with oxazole to give the coupled product (**7.57.**), consisting solely of the 5-oxazolyl isomer.[77] Extension of the reaction to imidazole led to a similar observation,[78] and the expected 5-pyrazyl-imidazole derivative was isolated in acceptable yield. Reactions using thiophene as the masked olefin gave similar results. Under forcing conditions the 2,5-diarylation of furan was also observed.[77]

(7.57.)

The coupling can also be extended to condensed systems. Indole, benzofurane or benzothiophene were all coupled with chloropyrazine derivatives with varying success. For example indole and the 3,6-diisobutyl-pyrazine derivative gave the desired product in moderate yield (**7.58.**). The coupling of the diethyl-pyrazine derivative with benzofuran led to 54% isolated yield, while the analogous process using benzothiophene afforded the 2-pyrazyl-benzothiophene in 81%.[77]

(7.58.)

A unique reaction of allenylindium reagents, prepared from propargyl bromides and indium, and haloazines gives rise to the appropriate allenyl-azines. Although the reaction is not a Heck coupling, but a cross-coupling reaction, the aryl moiety is formally attached to a carbon-carbon multiple bond, therefore we mention it here.

5-Bromopyrimidine and 1-bromo-2-butyne reacted to give the butadienyl-pyrimidine derivative in 92% yield, while the sequential coupling of propargyl bromide and 1-bromo-2-butyne with 2,5-dibromopyridine gave the expected product in 65% (**7.59.**), the first allene moiety being introduced into the more reactive 2-position, and the second allene attached to position 5.[79]

(7.59.)

CO-insertion (carbonylative coupling)

Although the insertion of carbon monoxide into arylpaladium complexes at ambient pressure is well documented the analogous reaction of azinylpalladium complexes usually requires elevated pressures due to the electron deficient nature of these heterocycles. The reaction is commonly used to convert the haloazine into the appropriate carboxylic acid or hetaryl-ketone derivative.

The carbonylation of 2-chloropyridine at high temperature in methanol under 4 bar CO pressure in the presence of a palladium-triphenylphosphine catalyst, for example, resulted in the formation of picolinic acid methyl ester in good yield (**7.60.**).[80] The equivalent amount of acid produced in the coupling is quenched by the added triethylamine. Interestingly, if the chlorine is in the less activated 3-position, the reaction fails. Bromo-, and iodopyridines undergo carbonylative coupling readily, irrespective of the position of the halogen in the ring.

$$\text{(7.60.)}$$

The carbonylation was extended to 2-chloro-3,6-dimethylpyrazine too (**7.61.**). Under the same conditions as in **7.60.** this compound led to the formation of the pyrazinecarboxylic ester in high yield.[80]

The analogous reaction of the 2-chloropyrimidine derivative in **7.62.** was also run at elevated temperature under 15 bar CO pressure. Depending on the alcohol, which was either added in excess or used as solvent, the desired esters were isolated in good to excellent yield. If the reaction was run at decreased carbon monoxide pressure, then the dehalogenation of the pyrimidine also became significant.[81] The effect of the used ligand was also tested and 1,1'-bis(diphenylphosphino)ferrocene (dppf) gave the best results.

$$\text{(7.61.)}$$

$$\text{(7.62.)}$$

The carbonylation in certain cases might not stop after the insertion of the first molecule of carbon monoxide and glyoxylic acid derivatives might be formed through double carbonylation. 2-, and 4-iodopyridine were reacted with butanol or diethylamine under a CO pressure of 60-90 bar and in most cases the pyridylglyoxylate ester or amide was formed with good selectivity and in good yield (**7.63.**). The side product of the reaction is the corresponding picolinic acid derivative.[82]

$$\text{(7.63.)}$$

Dihaloazines also undergo carbonylation and, if the positional activation of the halogens is different, then the reaction might be carried out selectively. 2,3-Dichloro-5-(methoxymethyl)pyridine was, for example, selectively carbonylated in the 2-position in methanol in the presence of a bis(diphenylphosphino)palladium dichloride, dppb catalyst system under 15 atm CO (**7.64.**).[83] The side product, present in less than 3% was the dicarbonylated derivative. A slight increase of the reaction temperature and the change of the catalyst system to palladium acetate and dppf, on the other hand, led to the predominant formation of the pyridinedicarboxylate diester in excellent yield.

(7.64.)

The inherent difference in the reactivity of the 2-, and 3-positions of the pyridine ring was also exploited in an industrial application of the carbonylative coupling of pyridines. 2,5-Dibromo-3-methylpyridine was converted into the 2-monoamides using different amines with a 98:2 selectivity. Keys to the success of the coupling, which was run on the 100 kg scale, were the use of DBU as base and the replacement of the phosphine in the catalyst with 2,2'-bipyridine.[84] The carbonylation of 3,5-dibromopyridine is less selective, although it could be converted into the pyridinedicarboxylic acid diester in quantitative yield.[85]

The selective carbonylation of 4,7-dichloroquinoline proceeds readily and the chlorine in the more active 4-position reacts preferentially to give quinolinecarboxylic acid derivatives in excellent yield (**7.65.**). The carbon-chlorine bond in the 7-position is so unreactive that attempts at the preparation of dicarboxylic acid derivatives were only moderately successful.[86]

(7.65.)

The carbonylative cross-coupling reactions of haloazines are usually run under an ambient to moderate carbon monoxide pressure. Arylboronic acids or tetraarylborates are usually the reagent of choice due to their robustness and availability. The coupling of 2-iodopyridine and phenylboronic acid under ambient CO pressure, for example, led to the formation of 2-benzoylpyridine in good yield (**7.66.**).[87]

In the analogous reaction of 2,5-dihalopyridines the formation of three different products was observed, their ratio depending on the applied conditions. At 80 °C and 5 bar CO pressure, starting from 5-bromo-2-

iodopyridine the predominant product was 2-benzoyl-5-brompyridine (**7.67.**) formed with 90% selectivity and isolated in 78% yield. Starting from 2,5-dibromopyridine and increasing the pressure to 50 bar and the temperature to 120 °C led to complete conversion in the 5-position too on prolonged heating. The product mixture in this case contains predominantly 2,5-dibenzoylpyridine (67%) along with some 2-benzoyl-5-phenylpyridine (not isolated). The reactions were run in the presence of the highly active bis(tricyclohexylphosphino)palladium dichloride catalyst.[88]

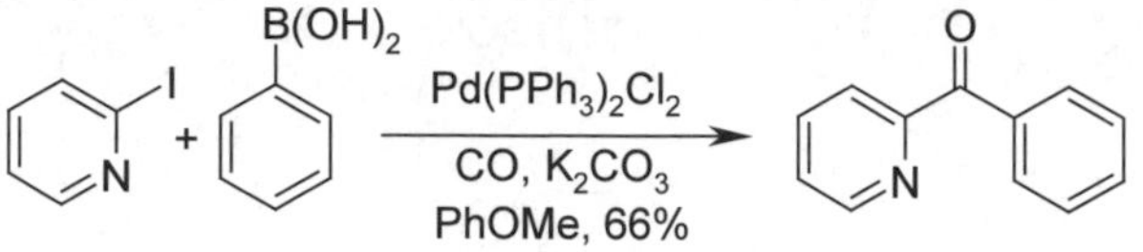

(7.66.)

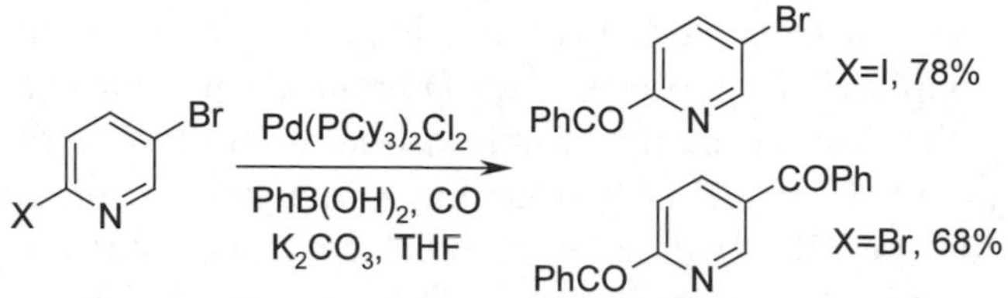

(7.67.)

The carbonylative cross-coupling was successfully extended to organofluorosilanes by Hiyama. *N,N'*-Dimethyl-2-imidazolidinone was found to be the most effective solvent for the carbonylative Hiyama-coupling, which was run in the presence of potassium fluoride. 3-Iodoquinoline, for example, reacted smoothly with 2-(ethyldifluorosilyl)-thiophene (**7.68.**) under an ambient pressure of carbon monoxide to give the desired ketone in 78% isolated yield.[89]

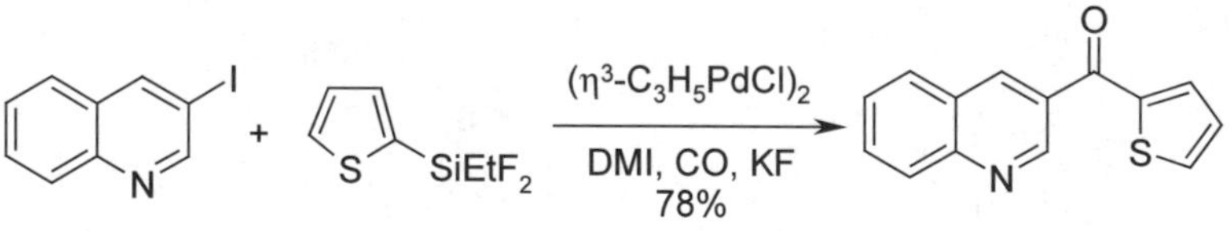

(7.68.)

7.3   CARBON-HETEROATOM BOND FORMATION

The introduction of nucleophiles onto haloazines in nucleophilic substitution has long been known, although most processes required forcing conditions. The palladium, nickel and copper-catalyzed carbon-heteroatom bond forming reactions usually provide a milder alternative to the original procedures.

2-Chloropyridine, for example, was found to react readily with cyclohexylamine already at 70 °C in the presence of a palladium-BINAP catalyst and sodium *tert*-butoxide (**7.69.**). Under similar conditions 2,6-dibromopyridine also reacted smoothly with the less nucleophilic aniline and the 2,6-disubstituted pyridine derivative was isolated in good yield (**7.70.**)[90]

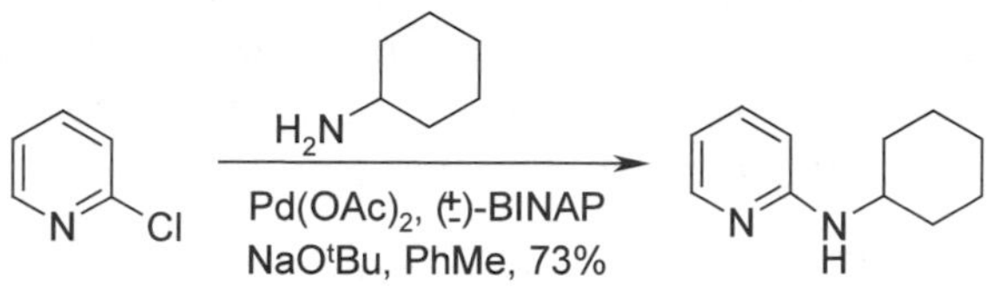

(7.69.)

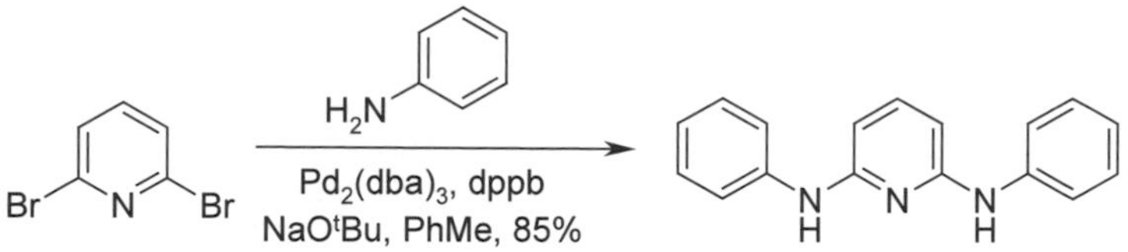

(7.70.)

The coupling of halopyridines was also extended to chiral amines. 2-Bromo-4-picoline was coupled with a series of enantiopure amines in the presence of Buchwald's palladium-BINAP catalyst system (**7.71.**).[91]

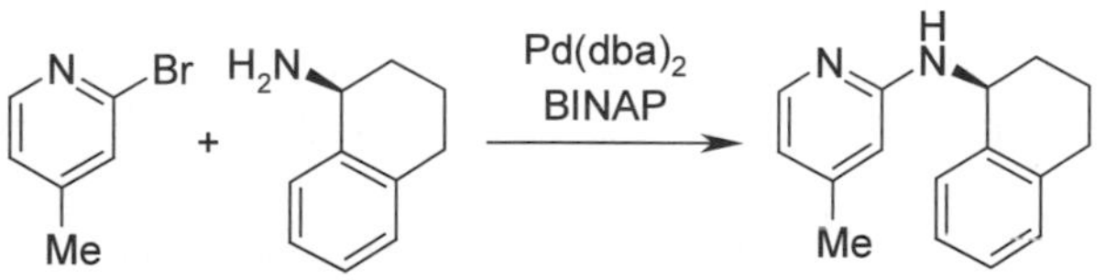

(7.71.)

Since the use of ammonia is not practical in transition metal catalyzed processes, the identification of its synthetic equivalents is of major importance. Benzophenone imine was found to couple with 3-bromopyridine readily under the above mentioned conditions (**7.72.**). The masking benzophenone was removed in transamination with hydroxylamine, which gave the desired 3-aminopyridine in 81% overall yield.[92] Allylamine was also successfully employed as ammonia equivalent.[93]

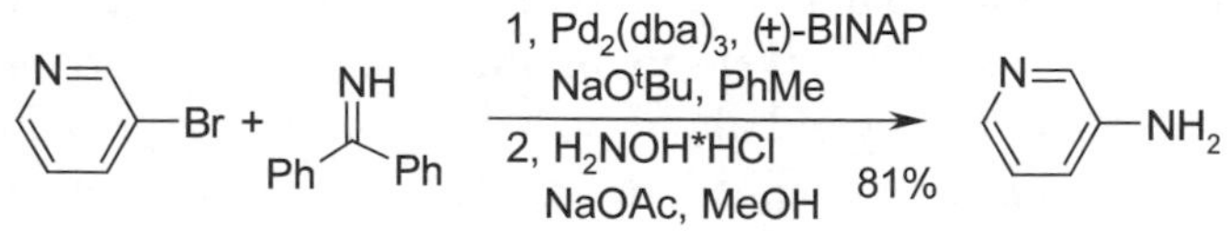

(7.72.)

An alternate approach, which utilizes lithium amides such as lithium hexamethyldisilazide or lithium amide, was also efficient in converting 2-chloropyridine into 2-aminopyridine (**7.73.**). In these reactions 2-(dicyclohexylphosphino)biphenyl was used as catalyst and the silyl protecting group was removed by TBAF.[94]

The use of the same, highly active catalyst system and microwave heating also allowed for the drastic reduction of the reaction time. The coupling of 3-chloropyridine and 4-toluidine in the presence of 1 % catalyst and sodium *tert*-butoxide gave on 10 minute irradiation the coupled product in 89% yield (**7.74.**). 2-Chloropyridine, 2-chloroquinoline and 2-chloropyrazine coupled equally well under the same conditions.[95]

(7.73.)

(7.74.)

Nickel catalysts are also efficient in promoting the coupling of halopyridines and amines. In the presence of nickel acetylacetonate and Arduengo type carbene precursors chloropyridines were coupled with a series of amines to give the aminopyridines in high yield. 3-Chloropyridine, for example, reacted with *N*-methylaniline to yield the anilinopyridine in 93% (**7.75.**). Of the carbene precursors tested 1,3-bis(2',6'-diisopropylphenyl)-4,5-dihydroimidazolium chloride was the most efficient.[96]

(7.75.)

2,2'-Bipyridine was also en efficient ligand for the nickel-catalyzed carbon-nitrogen bond forming reactions. Using piperazine as a bifunctional amine, its monoarylation and diarylation were both achieved selectively with 2-chloropyridine. By using a 1:1 ratio of the reagents the monoarylated product was isolated in 63% yield along with 12% of the diarylated product (**7.76.**). When the ratio was changed to 1:2 *N,N'*-dipyridylpiperazine was isolated in 78% yield along with 7% of the monocoupled product.[97]

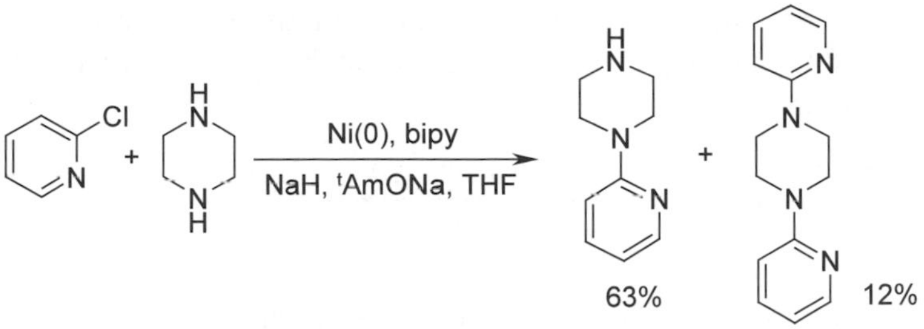

(7.76.)

In certain cases, when the palladium or nickel catalyzed coupling is not efficient or fails completely, an alternate solution is provided by the use of copper based catalyst systems. The 5-iodouracil derivative shown in **7.77.** was unreactive towards imidazole using either the Buchwald-Hartwig conditions or the copper(I) triflate promoted the carbon-nitrogen bond formation reported by Buchwald.[98] These latter conditions, however, were effective in coupling the iodouracil with a series of other amines (**7.77.**). The optimal catalyst system consisted of copper(I) triflate, phenantroline and dibenzylideneacetone (dba).[99]

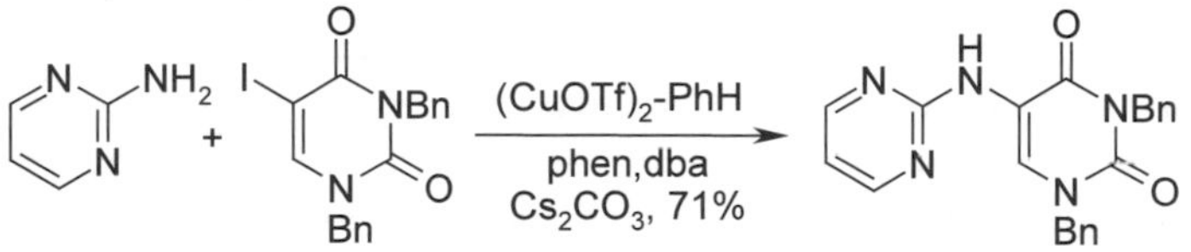

(7.77.)

Azines, and pyridine in particular, might also participate in coupling reactions through their ring nitrogen atom. 2-Hydroxylpyridine was found to react readily with aryl iodides through its tautomeric pyridone form in the presence of a copper-diamine catalyst and potassium phosphate, which acted in the process as base (**7.78.**). Different 2-pyridone derivatives were coupled efficiently giving the *N*-arylated products in good to excellent yield.[100] Strong electron withdrawing groups on the pyridine and *ortho*-substituents on either reagent were found to impede the coupling reaction.

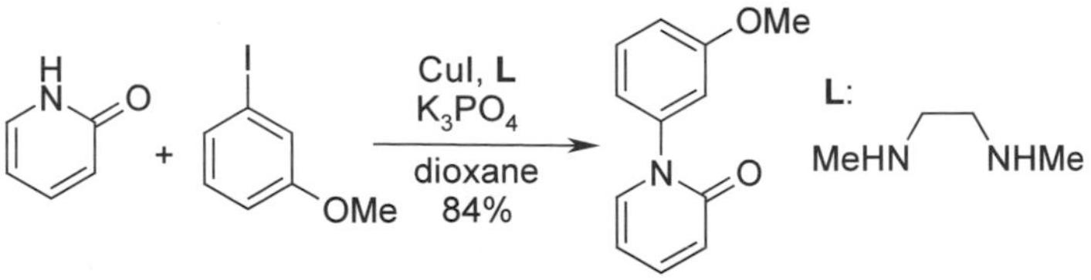

(7.78.)

The copper catalyzed carbon-heteroatom bond forming reactions are also efficient in the introduction of oxygen and phosphorous based substitutents onto the aromatic ring. 3-Iodopyridine was reacted with *n*-butanol in the presence of 10 mol% copper(I) iodide and 20 mol% 1,10-phenantroline using caesium carbonate as base, under which conditions 3-butoxypyridine was formed in 87% yield (**7.79.**). Secondary alcohols, such as isopropanol gave similar results.[101]

$$(7.79.)$$

The coupling of secondary phosphines with aryl iodides was also extended to pyridines. 3-Iodopyridine and diphenylphosphine were coupled to give 3-diphenylphosphino-pyridine in 60% yield (**7.80.**). The catalyst in this reaction consisted of copper(I) iodide and *N,N'*-dimethylethylenediamine and caesium carbonate was used again as base.[102]

$$(7.80.)$$

Finally, a recently reported copper catalyzed carbon-nitrogen bond forming process utilises reagents with polarity opposite to the common disconnection protocols. An electrophilic nitrogen, in most cases an *O*-acyl hydroxylamine derivative, was successfully coupled with diarylzinc reagents in the presence of copper triflate or copper chloride. Di(2'-pyridyl)zinc and *N*-benzoyloxy-morpholine were reacted at ambient temperature in the presence of 1% copper(I) triflate to give 2-morpholinopyridine in 71% yield (**7.81.**). Under these mild conditions the reaction was over in less than one hour.[103]

$$(7.81.)$$

7.4   OTHER PROCESSES

The transformations mentioned in this chapter either fit only loosely into the previously described categories, or could be listed in several of them. They usually have significant synthetic value or are of mechanistic interest.

The scope of the copper promoted nucleophilic displacement reactions on heterocyclic systems is not limited to nitrogen, oxygen and phosphorous nucleophiles. Buchwald and co-workers demonstrated that the same catalyst system that is efficient in carbon-phosphorous bond formation is also the catalyst of choice for the introduction of the cyano group onto the azine core. The process works well with the so far unreactive hetaryl bromides if a catalytic amount of potassium iodide is present. 3-Bromoquinoline, for example gave 3-cyanoquinoline in 78% isolated yield (**7.82.**). The *in situ* formation of the appropriate aryl iodide was proved experimentally.[104]

(7.82.)

This observation opens up the way for another transformation: the copper catalyzed halogen exchange reaction on aromatic systems. When not only a catalytic, but a stoichiometric amount of iodide was added to 3-bromoquinoline in the presence of the copper catalyst system (**7.83.**), it was converted smoothly to 3-iodoquinoline, which was isolated from the reaction mixture in 95% yield.[105]

(7.83.)

The coupling of aryl halides and classical carbon nucleophiles, such as malonates, is also feasible in the presence of a properly selected palladium or copper catalyst. Diethyl malonate and 3-iodopyridine, for example, gave diethyl 2-(3'-pyridyl)malonate in 73% yield (**7.84.**). The optimal catalyst in this process consisted of copper(I) iodide and 2-hydroxylbiphenyl.[106]

(7.84.)

The coupling of aromatic moieties sometimes can be achieved through the activation of a carbon-hydrogen bond. If such a reaction proceeds selectively, then the procedure is usually of great synthetic value due to its increased economy. Hetaryl halides are known to undergo such coupling with electron rich aromatic systems, such as azoles. The 2-chloropyrazine derivative, shown in **7.85.** was reacted with *N*-substituted indole derivatives in the presence of tetrakis(triphenylphosphino)palladium and potassium acetate. The pyrazinyl moiety was introduced into the 2- or 3-position depending on the steric bulk of the *N*-substituent.[107] The reactions might proceed through the electrophilic attack of the arylpalladium species on the indole, although in similar processes another mechanism was also found to intervene.[108]

It is interesting to point out, that the analogous reaction of 1,4-dichlorophthalazine and 1-methylindole (**7.86.**), when run under classical Friedel-Crafts conditions in the presence of AlCl$_3$ in dichloroethane, resulted in the selective formation of the 3-hetarylindole derivative in good yield.[109]

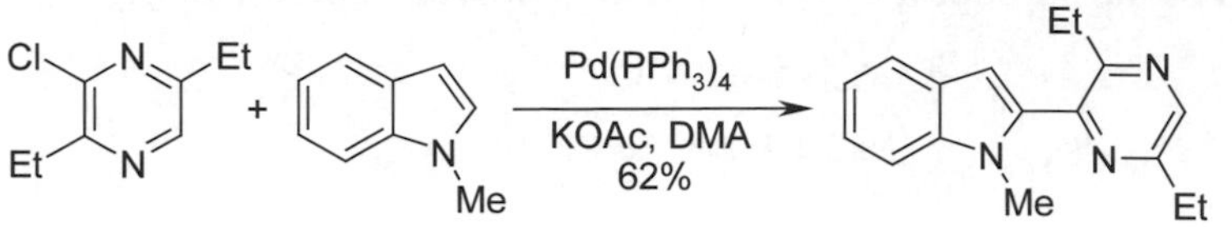

(7.85.)

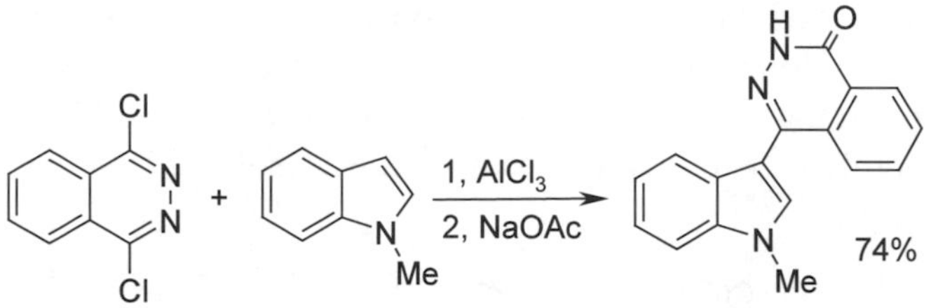

(7.86.)

When 4-methylpyrimidine was used as the coupling partner of bromobenzene (**7.87.**), the attack of the phenylpalladium intermediate took place rather at the relatively acidic methyl substituent than on the electron deficient pyrimidine core, giving rise to 4-diphenylmethyl-pyrimidine.[110]

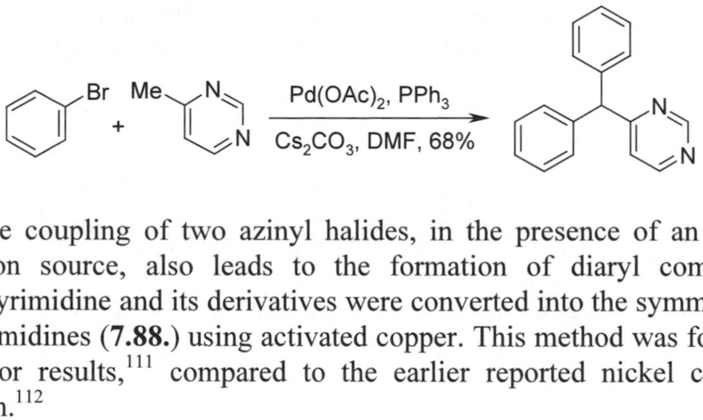

(7.87.)

The coupling of two azinyl halides, in the presence of an appropriate electron source, also leads to the formation of diaryl compounds. 2-Iodopyrimidine and its derivatives were converted into the symmetrical 2,2'-bipyrimidines (**7.88.**) using activated copper. This method was found to give superior results,[111] compared to the earlier reported nickel chloride-zinc system.[112]

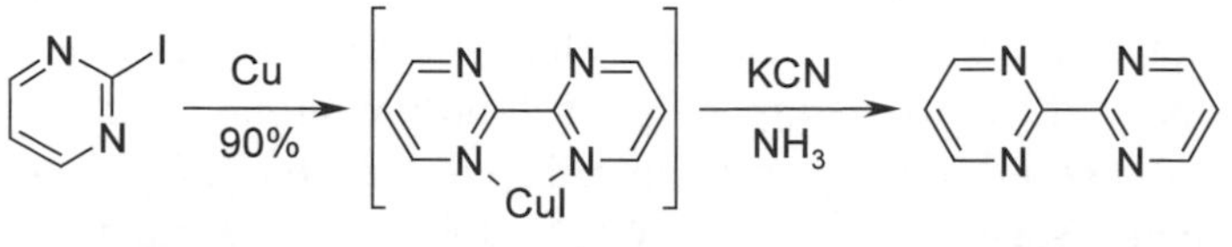

(7.88.)

Electrochemical processes provide a low waste alternative to the use of stoichiometric amounts of reducing agents. The homocoupling of 2-

bromopyridine derivatives was observed in the presence of a nickel(II) bromide – bipyridine catalyst system (**7.89.**). The proposed mechanism is a combination of oxidative addition and one electron reduction steps starting from a nickel(0) complex and concluding in the reductive elimination of bipyridine from a dipyridylnickel(II) complex.[113]

$$
\text{R} \diagup \underset{N}{\diagup} \text{Br} \quad \xrightarrow[\substack{\text{electrolysis} \\ 23\text{-}69\%}]{\text{NiBr}_2,\ \text{bipy}} \quad \text{R}\diagup\underset{N}{\diagdown}\diagup\underset{N}{\diagdown}\diagup\text{R} \qquad \text{R = H, CH}_3
$$

$$(7.89.)$$

The reagent that is introduced onto the heteroyclic core, in certain cases might already be part of the starting molecule. Such an intramolecular palladium-catalyzed rearrangement is depicted in **7.90**. Allyl groups were introduced into the 1-position of pyridine via the rearrangement of 2-allyloxy-pyridine derivatives.[114] The transformation was catalyzed by both palladium(II) and palladium(0) complexes, although they usually gave different products. In the former case the reaction goes probably through a cyclic intermediate, why in the latter case the process probably follows an oxidative addition – reductive elimination pathway.

cat:	PdCl$_2$	100%	0% Me
	Pd(PPh$_3$)$_4$	33%	67%

$$(7.90.)$$

7.5   REFERENCES

[1] (a) Miyaura, N.; Suzuki, A. *Chem. Rev.* **1995**, *95*, 2457. (b) Stanforth, S. S. *Tetrahedron* **1998**, *54*, 263. (c) Suzuki, A. *J. Organomet. Chem.* **1999**, *576*, 147.

[2] For a recent review on heterocyclic boronic acids see Tyrrel, E.; Brookes, P. *Synthesis*, **2003**, 469.

[3] Zoltewicz, J. A.; Cruskie, M. P., Jr. *Tetrahedron* **1995**, *51*, 11393.

[4] Li, J. J.; Yue, W. S. *Tetrahedron Letters* **1999**, *40*, 4507.

[5] (a) Ishikura, M.; Kamada, M.; Terashima, M. *Heterocycles* **1984**, *22*, 265. (b) Ishikura, M.; Kamada, M.; Terashima, M. *Synthesis* **1984**, 936.

[6] Deshayes, K.; Broene, R. D.; Chao, I.; Knobler, C. B.; Diederich, F. D. *J. Org. Chem.* **1991**, *56*, 6787.

[7] Parry, P. R.; Changsheng, W.; Batsanov, A. S.; Bryce, M. R.; Tarbit, B. *J. Org. Chem.* **2002**, *67*, 7541.

[8] Ishikura, M.; Oda, I.; Terashima, M. *Heterocycles* **1985**, *23*, 2375.

[9] Ishiyama, T.; Ishida, K.; Miyaura, N. *Tetrahedron*, **2001**, *57*, 9813.

[10] Fuller, A. A.; Hester, H. R.; Salo, E. V.; Stevens, E. P. *Tetrahedron Lett.* **2003**, *44*, 2935.

[11] Csányi, D.; Timári, G. Hajós, Gy. *Synth. Commun.* **1999**, *29*, 3959.

[12] Li, G. J. *J. Org. Chem.* **2002**, *67*, 3643 and references therein.

[13] Yang, W.; Wang, Y.; Corte, J. R. *Org. Lett.* **2003**, *5*, 3131.

[14] Schultheiss, N.; Bosch, E. *Heterocycles* **2003**, *60*, 1891.

[15] Bourotte, M.; Pellegrini, N.; Schmitt, M.;Bourguignon, J-J. *Synlett*, **2003**, 1482.

[16] Krajsovszky, G.; Mátyus, P.; Riedl, Zs.; Csányi, D.; Hajós, Gy. *Heterocycles* **2001**, *58*, 1105.

[17] Gronowitz, S.; Hoernfeldt, A. B.; Kristjansson, V.; Musil, T. *Chemica Scripta* **1986**, *26*, 305.

[18] Piettre, S. R.; Andre, C.; Chanal, M-C.; Ducep, J-B.; Lesur, B.; Piriou, F.; Raboisson, P.; Rondeau, J-M.; Schelcher, C.; Zimmermann, P.; Ganzhorn, A. J. *J. Med. Chem.* **1997**, *40*, 4208.

[19] Wellmar, U.; Hörnfeldt, A.-B.; Gronowitz, S. *J. Heterocycl. Chem.* **1995**, *32*, 1159.

[20] Kusturin, C.; Liebeskind, L. S.; Rahman, H.; Sample, K.; Schweitzer, B.; Srogl, J.; Neumann, W. L. *Org. Lett.* **2003**, *5*, 4349.

[21] Morris, G. A.; Nguyen, SB. T. *Tetrahedron Lett.* **2001**, *42*, 2093.

[22] Stanforth, S. P. *Tetrahedron* **1998**, *54*, 263 and references therein.

[23] Böhm, V. P. W.; Weskamp, T.; Gstöttmayr, C. W. K.; Herrmann, W. A. *Angew. Chem. Int. Ed.* **2000**, *39*, 1602.

[24] Mongin, F.; Mojovic, L.; Guillamet, B.; Trécourt, F.; Quéguiner, G. *J. Org. Chem.* **2002**, *67*, 8991.

[25] Korn, T. J.; Cahiez, G.; Knochel, P. *Synlett* **2003**, 1892.

[26] Anctil, E. J-G.; Snieckus, V. *J. Organomet. Chem.* **2002**, *653*, 150.

[27] Fang, Y-Q.; Hanan, G. S. *Synlett* **2003**, 852.

[28] Lutzen, A.; Hapke, M.; *Eur. J. Org. Chem.* **2002**, 2292.

[29] Stanetty, P.; Schnürch, M.; Mihovilovic, M. D. *Synlett* **2003**, 1862.

[30] Mongin, F.; Trécourt, F.; Gervais, B.; Mongin, O.; Quéguiner, G. *J. Org. Chem.* **2002**, *67*, 3272.

[31] Littke, A. F.; Schwarz, L.; Fu, G. C. *J. Am. Chem. Soc.* **2002**, *124*, 6343.

[32] Wolf, C.; Lerebours, R. *J. Org. Chem.* **2003**, *68*, 7077.

[33] For an example, 2-tributylstannylpyridine see Peters, D.; Hörnfeldt, A.-B.; Gronowitz, S. *J. Heterocycl. Chem.* **1990**, *27*, 2165.

[34] For a representative example see Benaglia, M; Ponzini, F.; Woods, C. R.; Siegel, J. S. *Org. Lett.* **2001**, *3*, 967.

[35] Schwab, P. F.; Fleischer, F.; Michl, J. *J. Org. Chem.* **2002**, *67*, 443.

[36] Heller, M.; Schubert, U. S. *J. Org. Chem.* **2002**, *67*, 8269.

[37] Peters, D.; Hörnfeldt, A.-B.; Gronowitz, S. *J. Heterocycl. Chem.* **1990**, *27*, 2165.

[38] Yamamoto, Y.; Tanaka, T.; Yagi, M.; Inamoto, M. *Heterocycles*, **1996**, *42*, 189.

[39] Takahashi, T.; Koga, H.; Sato, H.; Ishizawa, T.; Taka, N. *Heterocycles*, **1995**, *41*, 2405.

[40] Barros, M. T.; Maycock, C. D.; Ventura, M. R. *Tetrahedron Lett.* **1999**, *40*, 557.

[41] Kennedy, G.; Perboni, A. D. *Tetrahedron Lett.* **1996**, *37*, 7611.

[42] Le Floch, P.; Carmichael, D.; Ricard, L.; Mathey, F. *J. Am. Chem. Soc.* **1993**, *115*, 10665.

[43] (a) Minn, K. *Synlett* **1991**, 115. (b) Ames, D. E.; Bull, D.; Takunda, C. *Synthesis* **1981**, 364.

[44] Louërat, F.; Gros, P.; Fort, Y. *Tetrahedron Lett.* **2003**, *44*, 5613.

[45] Erdélyi, M.; Gogoll, A. *J. Org. Chem.* **2001**, *66*, 4165.

[46] Kim, C.-S.; Russel, K. C. *J. Org. Chem.* **1998**, *63*, 8229.

[47] Takahashi, S.; Kuroyama, Y.; Sonogashira, K.; Higihara, N. *Synthesis* **1980**, 627.

[48] (a) Novák, Z.; Nemes, P.; Kotschy, A. *Org. Lett.* **2004**, *6*, 4917. (b) For a review on domino Sonogashira coupling see Nagy, A.; Novák, Z.; Kotschy, A. *J. Organomet. Chem.* **2005**, in press.

[49] Novák, Z.; Szabó, A.; Répási, J.; Kotschy, A. *J. Org. Chem.* **2003**, *68*, 3327.

[50] Mori, Y.; Seki, M. *J. Org. Chem.* **2003**, *68*, 1571.

[51] Akita, Y.; Ohta, A. *Heterocycles* **1982**, *19*, 329.

[52] Edo, K.;Sakamoto, T.; Yamanaka, H. *Chem. Pharm. Bull.* **1978**, *26*, 3843.

[53] Napoletano, M.; Norcini, G.; Pellacini, F.; Marchini, F.; Morazzoni, G.; Ferlenga, P.; Pradella, L. *Bioorganic & Medicinal Chemistry Letters* **2001**, *11*, 33.

[54] Novák, Z.; Kotschy, A. *Org. Lett.* **2003**, *5*, 3495.

[55] Faragó, J.; Novák, Z.; Schlosser, G.; Csámpai, A.; Kotschy, A. *Tetrahedron*, **2004**, *60*, 1991.

[56] Samaritani, S.; Menicagli, R. *Tetrahedron* **2002**, *58*, 1381.

[57] Sonoda, M.; Inaba, A.; Itahashi, K.; Tobe, Y. *Org. Lett.* **2001**, *2*, 2419.

[58] Pauvert, M.; Laine, P.; Jonas, M.; Wiest, O. *J. Org. Chem.* **2004**, *69*, 543.

[59] Gelman, D.; Tsvelikhovsky, D.; Molander, G. A.; Blum, J. *J. Org. Chem.* **2002**, *67*, 6287.

[60] Beletskaya, I. P.; Latyshev, G. V.; Tsvetkov, A. V.; Lukashev, N. V. *Tetrahedron Lett.* **2003**, *44*, 5011.

[61] Penney, J. M.; Miller, J. A. *Tetrahedron Lett.* **2004**, *45*, 4989.

[62] McElroy, W. T.; DeShong, P. *Org. Lett.* **2003**, *5*, 4779.

[63] Basu, B.; Freijd, T. *Acta Chem. Scand.* **1996**, *50*, 316.

[64] Niu, C.; Li, J.; Doyle, T. W.; Chen, S.-H. *Tetrahedron* **1998**, *54*, 6311.

[65] Draper, T. L.; Bailey, T. R. *Synlett* **1995**, 157.

[66] Wei, Z-L.; George, C.; Kozikowski, A. P. *Terahedron Lett.* **2003**, *44*, 3847.

[67] Raggon, J. W.; Snyder, W. M. *Org. Proc. Res. & Dev.* **2002**, *6*, 67.

[68] Karig, G.; Moon, M.-T.; Thasana, N.; Gallagher, T. *Org. Lett.* **2002**, *4*, 3115.

[69] Karig, G.; Thasana, N.; Gallagher, *Synlett* **2002**, 808.

[70] Frantz, R.; Granier, M.; Durand, J.-O.; Lanneau, G. F. *Tetrahedron Lett.* **2002**, *43*, 9115.

[71] Zhao, J.; Yang, X.; Jia, X.; Luo, S.; Zhai, H. *Tetrahedron* **2003**, *59*, 9379.

[72] Turner, S. C.; Zhai, H.; Rapoport, H. *J. Org. Chem.* **2000**, *65*, 861.

[73] Sakamoto, T; Arakida, H.; Edo, K.; Yamanka, H. Heterocycles **1981**, *16*, 965.

[74] Köhler, K.; Heidenreich, R. G.; Krauter, J. G. E.; Pietsch, J. *Chem. Eur. J.* **2002**, *8*, 622.

[75] Walker, J. A., II; Chen, J. J.; Hinkley, J. M.; Wise, D. S.; Townsend, L. B. *Nucleosides Nucleotides* **1997**, *16*, 1999.

[76] Ma, S.; Zhang, J.; Lu, L.; *Chem. Eur. J.,* **2003**, *9*, 2447.

[77] Aoyagi, Y.; Inoue, A.; Koizumi, I.; Hashimoto, R.; Tokunaga, K.; Gohma, K.; Komatsu, J.; Sekine, K.; Miyafuji, A.; Konoh, J.; Homna, R.; Akita, Y.; Ohta A. *Heterocycles* **1992**, *33*, 257.

[78] Yamashita, M.; Oda, M.; Hayashi, K.; Kawasaki, I.; Ohta, S. Heterocycles, **1998**, 48, 2543

[79] Lee, K.; Seomoon, D.; Lee, P. H. *Angew. Chem. Int. Ed.* **2002**, *41*, 3901.

[80] Takeuchi, R.;Suzuki, K.; Sato, N. *Synthesis* **1990**, 923.

[81] Bessard, Y.; Crettaz, R. *Tetrahedron* **1999**, *55*, 405.

[82] Couve-Bonnaire, S.; Carpentier, J.-F.; Castanet, Y.; Mortreux, A. *Tetrahedron Lett.* **1999**, *40*, 3717.

[83] Bessard, Y.; Roduit, J. P. *Tetrahedron* **1999**, *55*, 393.

[84] Wu, G. G.; Wong, YS; Poirier, M. *Org. Lett.* **1999**, *1*, 745.

[85] Head, R.A.; Ibbotson, A. *Tetrahedron Lett.* **1984**, *25*, 5953.

[86] Najiba, D.; Carpentier, J.-F.; Castanet, Y.; Biot, C.; Brocard, J.; Mortreux, A. *Tetrahedron Lett.* **1999**, *40*, 3719.

[87] Ishiyama, T.; Kizaki, H.; Miyaura, N.; Suzuki, A. *Tetrahedron Lett.* **1993**, *34*, 2127.
[88] Couve-Bonnaire, S.; Carpentier, J.-F.; Mortreux, A.; Castanet, Y. *Tetrahedron* **2003**, *59*, 2793.
[89] Hatanaka, Y.; Fukushima, S.; Hiyama, T. *Tetrahedron* **1992**, *48*, 2113.
[90] Wagaw, S.; Buchwald, S.L. *J. Org. Chem.* **1996**, *61*, 7240.
[91] Rupert, K. C.; Henry, J. R.; Dodd, J. H.; Wadsworth, S. A.; Cavender, D. E.; Olini, G. C.; Fahmy, B.; Siekierka, J. J. *Bioorg.Med. Chem. Lett.* **2003**, *13*, 347.
[92] Wolfe, J. P.; Ahman, J.; Sadighi, J. P.; Buchwald, S. L. *Tetrahedron Lett.* **1997**, *38*, 6367.
[93] Jaime-Figueroa, S.; Liu, Y.; Muchowski, J. M.; Putnam, D. G. *Tetrahedron Lett.* **1998**, *39*, 1313.
[94] Huang, X.; Buchwald, S.L. *Org. Lett.* **2001**, *3*, 3417.
[95] Maes, B.U.W.; Loones, K.T.J.; Lemiére, G.L.F.; Dommisse, R.A. *Synlett* **2003**, 1822.
[96] Desmarets, C.; Schneider, R.; Fort, Y. *J. Org. Chem.* **2002**, *67*, 3029.
[97] Brenner, E.; Schneider, R.; Fort, Y. *Tetrahedron Lett.* **2000**, *41*, 2881.
[98] Kiyomori, A.; Marcoux, J. F.; Buchwald, S. L. *Tetrahedron Lett.* **1999**, *40*, 2657.
[99] Arterburn, J.B.; Pannala, M.;Gonzalez, A.M. *Tetrahedron Lett.* **2001**, *42*, 1475.
[100] Li, C.S.; Dixon, D.D. *Tetrahedron Lett.* **2004**, *45*, 4257.
[101] Wolter, M.; Nordmann, G.; Job, G.E.; Buchwald, S.L. *Org. Lett.* **2002**, *4*, 973.
[102] Gelman, D.; Jiang, L.; Buchwald, S.L. *Org. Lett.* **2003**, *5*, 2315.
[103] Berman, A.M.; Johnson, J.S. *J. Am. Chem. Soc.* **2004**, *126*, 5680.
[104] Zanon, J.; Klapars, A.; Buchwald, S.L. *J. Am. Cem. Soc.* **2003**, *125*, 2890.
[105] Klapars, A.; Buchwald, S.L. *J. Am. Chem. Soc.* **2002**, *124*, 14844.
[106] Hennessy, E.J.; Buchwald, S.L. *Org. Lett.* **2002**, *4*, 269.
[107] Akita, Y.; Itagaki, Y.; Takizawa, S.; Ohta, A. *Chem. Pharm. Bull.* **1989**, *37*, 1477.
[108] Wang, L.; Woods, K. W.; Li, Q.; Barr, K. J.; McCroskey, R. W.; Hannick, S. M.; Gherke, L.; Credo, R. B.; Hui, Y.-H.; Marsh, K.; Warner, R.; Lee, J. Y.; Zielinski-Mozng, N.; Frost, D.; Rosenberg, S. H.; Sham, H. L. *J. Med. Chem.* **2002**, *45*, 1697.
[109] Pal, M.; Batchu, V. R.; Parasuraman, K.; Yeleswarapu, K. R. *J. Org. Chem.* **2003**, *68*, 6806.
[110] Inoh, J.-I.; Satoh, T.; Pivsa-Art, S.; Miura, N.; Nomura, M. *Tetrahedron Lett.* **1998**, *39*, 4673.
[111] Vlád, G.; Horváth, I. T. *J. Org.Chem.* **2002**, *67*, 6550.
[112] Fort, Y.; Becker, S.; Caubére, P. *Tetrahedron* **1994**, *50*, 11893.
[113] De Franca, K. W. R.; Navarro, M.; Léonel, E.; Durandetti, M.; Nédélec, J.-Y. *J. Org. Chem.* **2002**, *67*, 1838.
[114] Itami, K.; Yamazaki, D.; Yoshida, J.-i. *Org. Lett.* **2003**, *5*, 2161

Chapter 8

THE FUNCTIONALIZATION OF OTHER RING SYSTEMS

This chapter discusses the transition metal catalyzed functionalization of such systems that fall outside the topic of Chapters 6 and 7, as well as certain other compound classes (*e.g.* purines, pyrones). In contrast to the abundant literature of the chemistry of five and six membered systems, the transition metal catalyzed transformations of other heterocycles have not been studied so far in the same depth, probably due to the limited availability of their halogen derivatives compared to haloazines and haloazoles. Purine compounds and their structural analogues constitute an exception, since their biological importance proved to be a strong drive for synthetic chemist worldwide.[1]

8.1   TRANSMETALATION ROUTE

As the examples will also illustrate, the vast majority of synthetic studies on the compounds discussed in this chapter were directed at their cross-coupling reactions. The target compounds were usually built up from the halogen derivative of the heterocycle, which was reacted with an organoelement compound, although examples where the roles of the halide and organometallic reagent were switched will also be presented. The featured reactions were divided into subclasses along the commonly used name reactions.

Suzuki coupling

The recent emergence of the Suzuki coupling as the method of choice for the cross-coupling of aryl halides,[2,3] due to its functional group tolerance and the stability of organoboron compounds, is clearly reflected by the proportion of these transformations in the literature.

Since the Suzuki coupling of purine derivatives was covered by recent reviews,[1] we only present a selection from these reactions. Xanthine (3,6-dihydropurine-2,6-dione) derivatives were coupled with different boronic acids, including styrylboronic acid, in the presence of the conventional tetrakis(triphenylphosphino)palladium catalyst and tripotassium phosphate as a "mild" base (**8.1.**), to obtain the appropriate 8-substituted xanthines in acceptable yield.[4] The advantage of the use of anhydrous tripotassium phosphate as base over the "classical" aqueous carbonate or hydroxide reagents might be attributed to the sensitivity of the 8-halopurine core towards nucleophilic attack.

(8.1.)

Hocek used Suzuki coupling and the iron catalyzed Grignard-coupling methodology, introduced by Fürstner,[5] to functionalise dihalopurine derivatives selectively. 2,6-Dichloropurines underwent cross-coupling in the 6-position with methylmagnesium chloride (**8.2.**) and the resulting 2-chloro-6-methylpurine was arylated under standard Suzuki coupling conditions to give the desired 2-aryl-6-methylpurines in good yield.[6] (*N.B.* The Suzuki coupling of phenylboronic acid on the same 2,6-dichloropurine does also proceed in the 6-position, due to its enhanced reactivity.[7])

(8.2.)

Surprisingly, when the same authors started from a 6,8-dichloropurine derivative, the course of the reaction depended on the reagent. Phenylboronic acid showed a marked preference for the 6-position again, while the iron catalyzed coupling of methylmagnesium chloride proceeded selectively in the 8-position (**8.3.**). The rationale behind the observed selectivity is still unclear.[8]

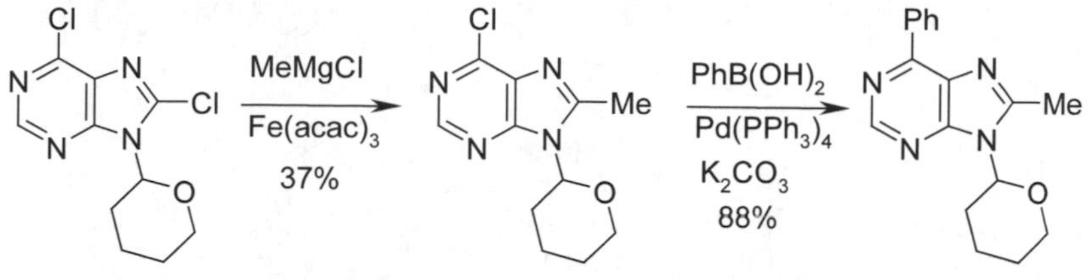

(8.3.)

The structural similarity of 6+5 condensed heteroaromatic ring systems, which contain nitrogen in both rings, to purines made these compounds a favourite target in pharmaceutical chemistry. The preparation of substituted pyrazolo[3,4-*d*]pyrimidines and pyrrolo[2,3-*d*]pyrimidines was achieved in a sequential nucleophilic displacement and Suzuki coupling (**8.4.**). An interesting feature of these transformations is that they were carried out in one pot and both were assisted (and greatly accelerated) by microwave irradiation.[9]

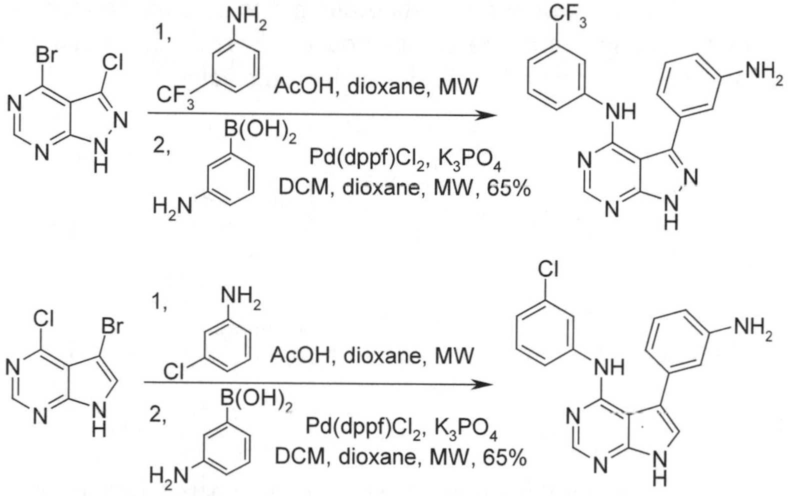

(8.4.)

In the search for effective kinase inhibitors 3-bromo-8-(3'-thienyl)-pyrazolo[2,3-*a*]pyrimidine, a positional isomer of the previously used system, was effectively coupled with a pyridineboronic ester *en route* to an active inhibitor (**8.5.**).[10]

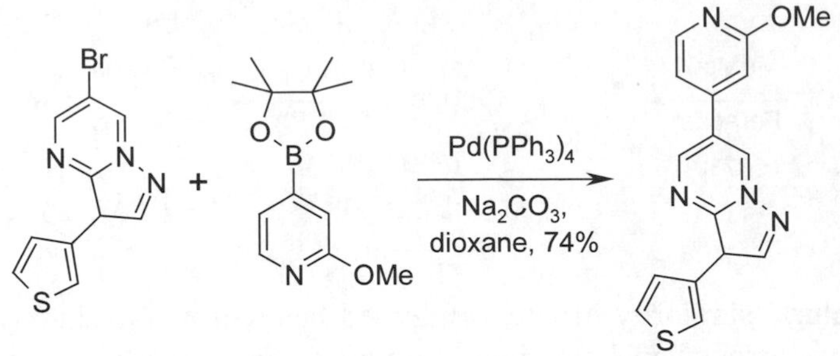

(8.5.)

The Suzuki coupling was also effective in the preparation of dimeric pyrrolo[2,3-*b*]pyridine derivatives, which serve as new melatonine analogues. The brominated azaindole shown in **8.6.** was converted to the corresponding boronate through a palladium catalyzed coupling with bis(dipinacolato)diborane. The formed hetarylboronic acid ester was coupled with another molecule of the starting material to give the desired bis-heterocycle. The two steps were run in the same pot to give an isolated yield of 83%.[11] The benzologue of the parent heterocycle, pyrrolo[2,3-*b*]quinoline, was also arylated efficiently with different arylboronic acids.[12]

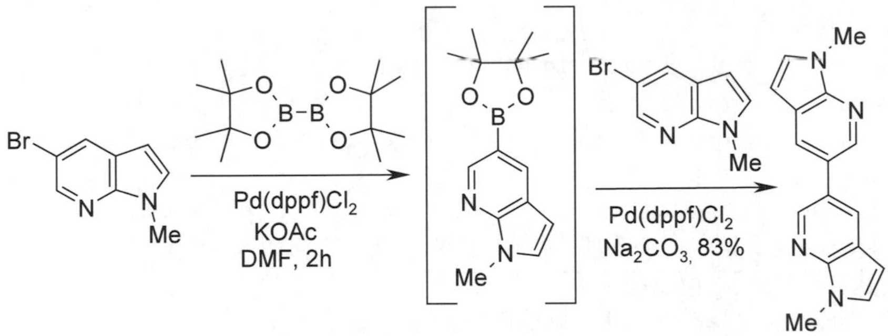

(8.6.)

An example of the participation of a seven membered heterocycle in Suzuki coupling is provided in equation **8.7.** The brominated azepino[4,3-*b*]indole derivative was reacted with a series of boronic acids, including the depicted 2-bezothiopheneboronic acid, to give the expected products in good yield. The coupling of vinylstannanes with the same azepinoindole derivative led to the introduction of olefins onto the heterocyclic core.[13] There are also a series of recent examples for the successful functionalization of the benzodiazepine system using Suzuki coupling reactions.[14,15]

Spiropyranes were also arylated efficiently using the Suzuki protocol. The reaction of diiodo-spiropyrane (**8.8.**) with an excess of phenylboronic acid led to the formation of the diarylated product in 91% yield. The analogous reaction starting from the dibromo compound was less effective and gave the desired product only in moderate yield. The synthesised spiropyrane derivatives were cleaved by ozonolysis to give salicylaldehydes.[16]

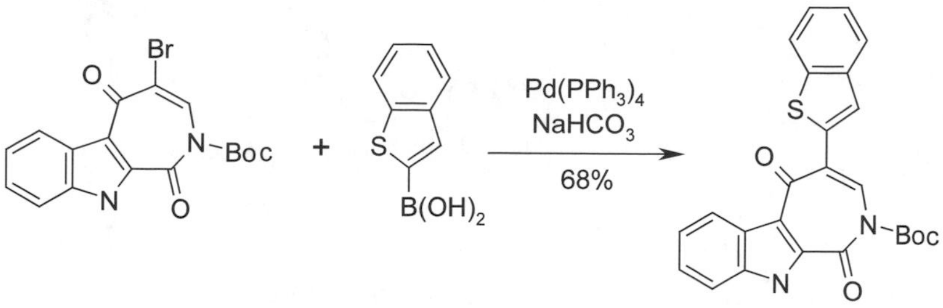

(8.7.)

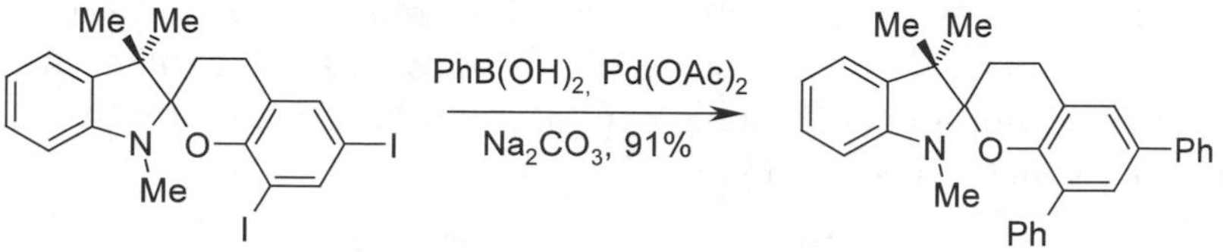

(8.8.)

The synthesis of pyrone derivatives attracted attention due to their synthetic potential. In an illustrative example Cho and co-workers studied the Suzuki-coupling of 3,5-dibromo-2-pyrone with arylboronic acids (**8.9.**). Under regular conditions the aryl group is introduced selectively into the more electron deficient 3-position, while in the presence of an equimolar amount of copper(I) iodide the coupling is diverted selectively into the 5-position (*N.B.* drop of the reaction temperature from 50 °C to ambient temperature negated the effect of copper and led to 3-arylation). The way copper effects the coupling is still unclear, but it was successfully used in the preparation of a range of 5-aryl-2-pyrons.[17]

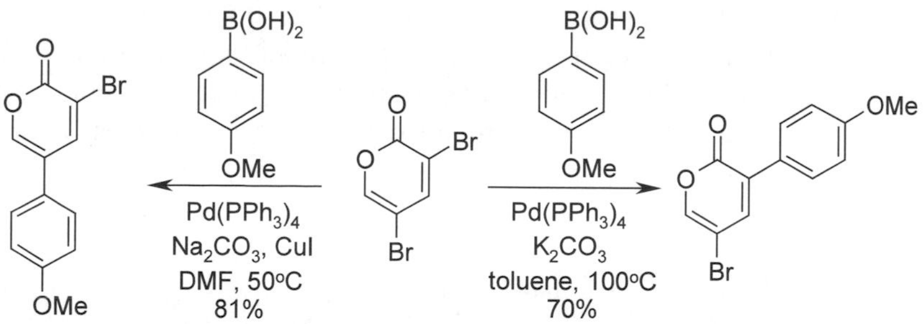

(8.9.)

4-Substituted benzo[*b*]pyran-2-ones (coumarins), as a privileged scaffold, exhibit interesting biological properties. The introduction of different aryl groups into the 4-position was accomplished by Fathi and Yang in Suzuki coupling (**8.10.**). The interesting feature of the coupling is that they utilized 4-tosyloxycoumarin as starting material. The willingness of the tosyl group to undergo oxidative addition might be attributed to the electron deficient nature of the 4-position.[18]

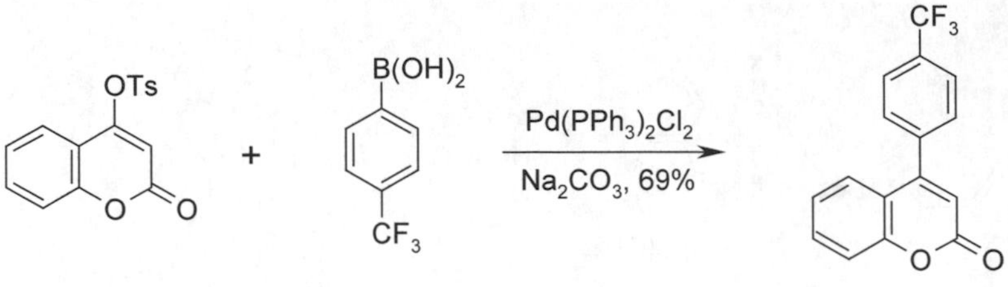

(8.10.)

The reactivity of bromopyrones can be increased through the depletion of their electron density *via* coordination to an electron withdrawing metal fragment. Fairlamb and co-workers converted 4-bromo-6-methyl-2-pyrone into its irontricarbonyl derivative using $Fe_2(CO)_9$ (**8.11.**), which underwent Suzuki coupling with 4-anisylboronic acid at room temperature to give the coupled complex in 64% isolated yield.[19]

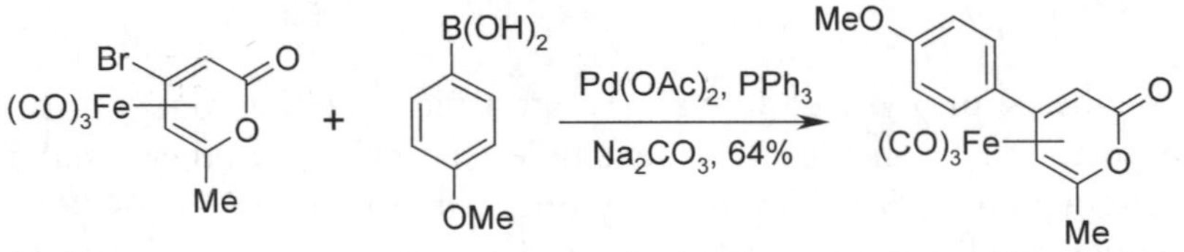

(8.11.)

Kharasch (Kumada) coupling

The palladium or nickel catalyzed cross-coupling of large and small ring heterocycles is not as common, as the Suzuki reaction. Except for a German patent, which describes the conversion of the chlorinated benzodiazepine derivative to the appropriate cyclohexyl compound using cyclohexylmagnesium chloride in the presence of a stoichiometric amount of manganese dichloride (**8.12.**),[20] other examples that should be mentioned here include the iron catalyzed transformations of halopurines using Grignard reagents, which were already discussed in **8.2.** and **8.3.**.[5-8]

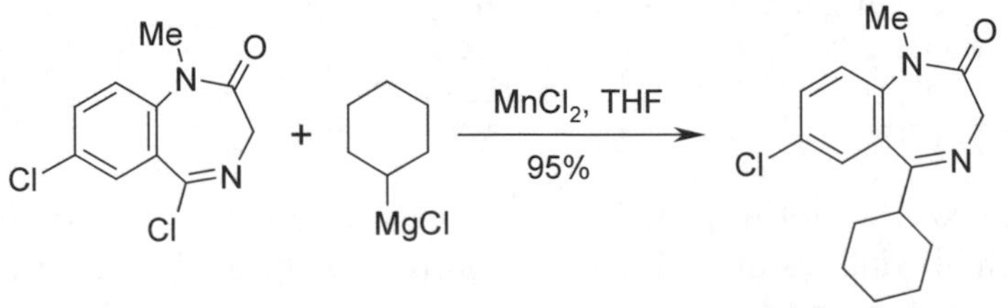

(8.12.)

Negishi coupling

The cross-coupling of organozinc reagents, which usually don't initiate unwanted side reactions like the Grignard reagents,[21] is well documented for purine derivatives and other related systems. 6-Iodo-9-tetrahydropyranyl-purine, for example, reacted readily with phenethylzinc chloride in the presence of a palladium-tripenylphosphine catalyst system to give 6-*β*-phenylethyl-purine in 74% yield after removal of the THP protecting group (**8.13.**).[22]

The cross-coupling of 9-benzyl-2,6-dichloropurine and methylzinc bromide showed good selectivity (**8.14.**). Since the enhanced reactivity of the 6-position already manifested in the Suzuki coupling of dichloropurines, it is not surprising that the protected 2-chloro-6-methylpurine is the major product in the process (71%) accompanied by some 9-benzyl-2,6-dimethylpurine (14%). In the light of the fact, that 1.2 equivalent of methylzinc bromide was used in the reported reaction, the process appears to be very efficient.[6] A similar result was obtained, when the organozinc reagent was generated *in situ* by the transmetalation of methylmagnesium chloride with zinc dibromide.[23] The reaction of an equimolar amount of *para*-methoxybenzylzinc chloride and 2,6-dichloro-9-isopropylpurine was also selective, yielding the 6-benzylated purine derivative in 71%.[24]

(8.13.)

(8.14.)

The selective sequential Suzuki-Negishi coupling of anisylboronic acid and *para*-methoxybenzylzinc chloride on 2,6-dichloro-9-isopropylpurine was also achieved (**8.15.**). Sequential addition of the coupling partners resulted in the selective introduction of the first aryl moiety into the 6-position followed by the introduction of the second coupling partner into the 2-position, leading to the isolation of a single product in excellent (91% yield).[24]

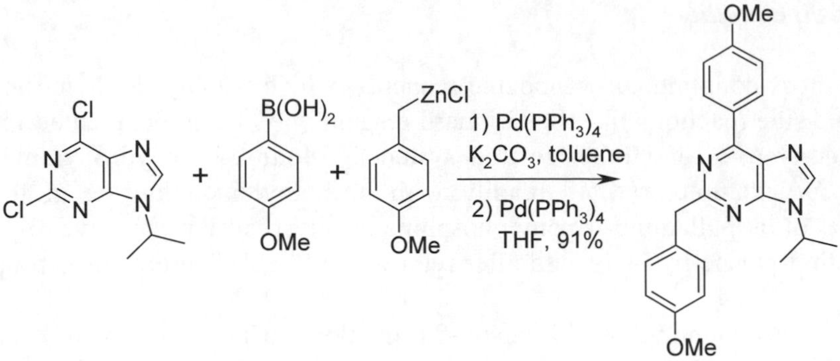

(8.15.)

The 5,7-dichloropyrazolo[1,5-*a*]pyrimidine system, a distant analogue of the above discussed dichloropurines, underwent Negishi coupling also in a selective manner (**8.16.**). Its reaction with a benzylzinc reagent led to the formation of the 7-benzylated isomer with high selectivity. Interestingly, addition of a stoichiometric amount of lithium chloride in place of the palladium catalyst reverses the selectivity and gives rise to the 5-benzylated product. The unreacted chlorine was also exchanged in the formed intermediate using phenylboronic acid and the Suzuki coupling protocol.[25]

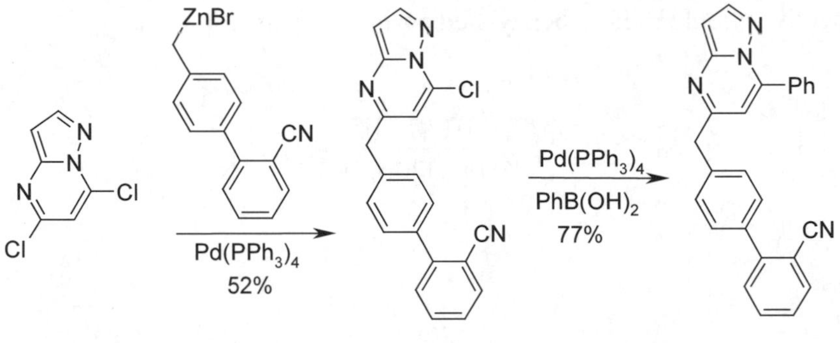

(8.16.)

The Negishi coupling was also effective in the functionalization of 1,4-benzodiazepine derivatives. The imidoyl chloride subunit of the 5-chloro-1,4-benzodiazepine derivative shown in **8.17.** was treated with an organozinc reagent derived from ethyl 6-bromohexanoate in the presence of bis(triphenylphosphineo)palladium dichloride in THF. The reaction afforded the alkylated benzodiazepine in good yield.[15]

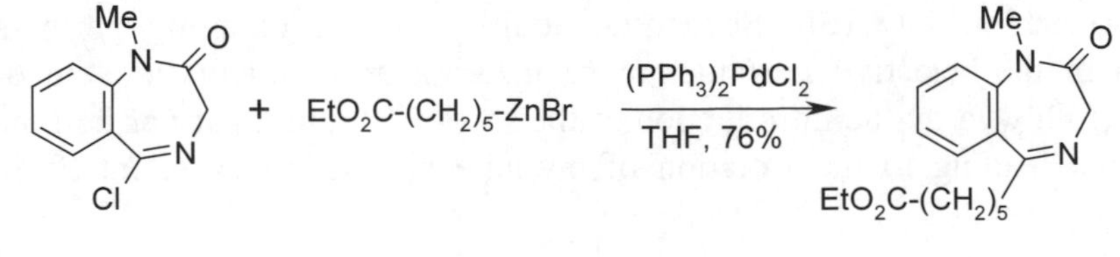

(8.17.)

Stille coupling

Organotin compounds, like their zinc and boron analogues are frequently employed in the functionalization of purine compounds. The cross-coupling of protected 2,6-dichloropurine with 2-tributylstannylfuran, for example, proceeded with excellent selectivity to give the 6-furylpurine derivative (**8.18.**).[26] The regioselectivity is in complete agreement with the previously discussed transformations of 2,6-dichloropurines.[6]

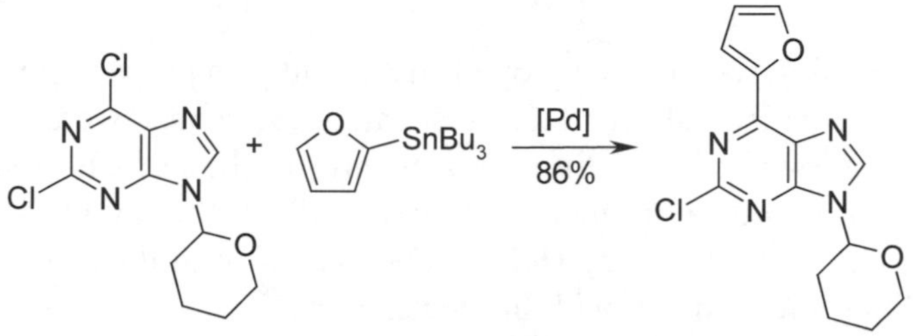

(8.18.)

The same selectivity was observed, when the coupling partner was changed to the styrylstannane shown in **8.19**. The styryl functionality was introduced into the 6-position exclusively with retention of its *E*-geometry. The same selectivity was observed with ethynylstannanes too. The resulting 2-chloropurine derivatives could be coupled further in the 2-position, which was exploited in a one-pot protocol leading to 2,6-disubstituted purine derivatives in a selective manner.[27]

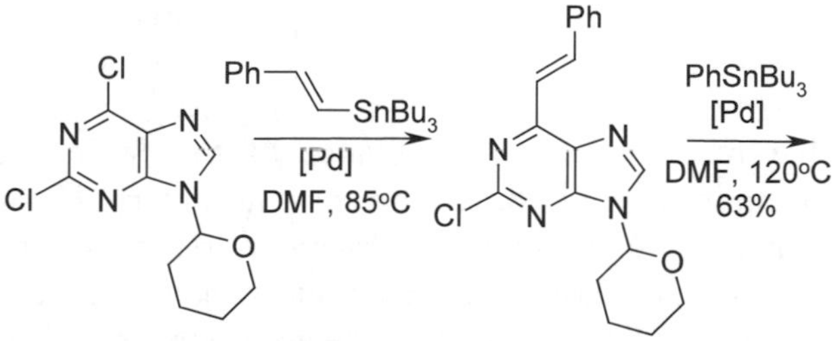

(8.19.)

Members of the she same compound class, arylethenylpurines, can also be prepared in a two step sequence. The cross-coupling of a halopurine with vinyl-tributylstannane leads to the formation of a vinylpurine, which in turn can undergo palladium catalyzed Heck reaction with a series of aryl halides (**8.20.**).[28] The two step procedure is of particular interest, since the alternate approach, the Heck reaction of halopurines and arylethenes is of very limited scope.

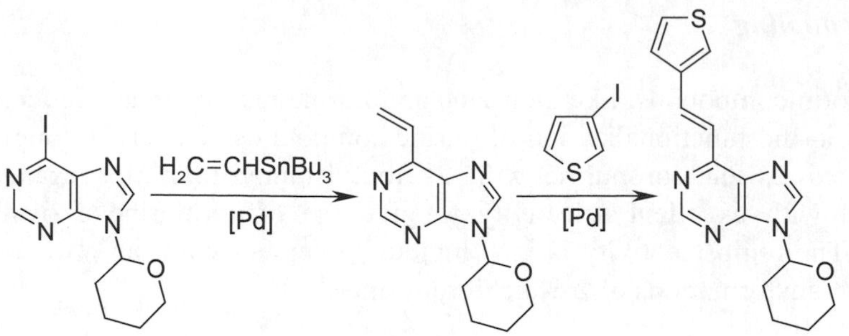

$$(8.20.)$$

The Stille coupling was also employed successfully to prepare covalently bound DNA base-pairs. 1,4-, and 1,3-bis(stannyl)benzenes were used to connect purine units covalently through a phenylene linker. The reactions gave the desired products in mediocre yield, along the by-products of destannylation and mono-coupling (**8.21.**). The extension of the procedure to benzene-1,4-diboronic acid met with limited success.[29]

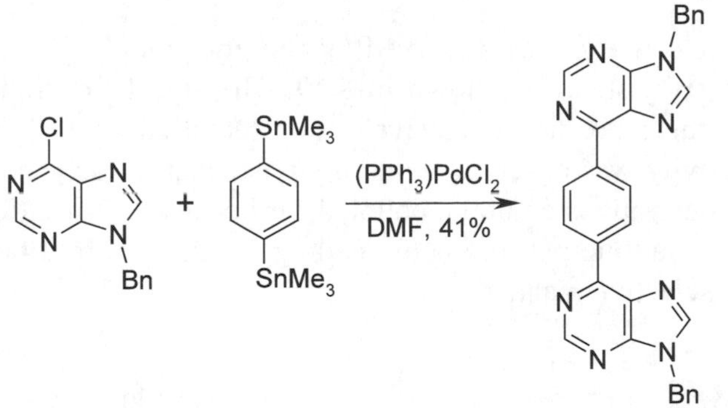

$$(8.21.)$$

The remarkable functional group tolerance of the Stille coupling is highlited in the example shown in **8.22**. The condensed benzo[1,4]diazepine derivative was reacted with a series of organotin reagents including 2-tributylstannyl-thiophene to isolate the coupling products in good yield. Interestingly, the attempted synthesis of a hetarylstannane through cross-coupling of the hetaryl triflate with hexamethyl-distannane gave only the reduction product in mediocre.[30,31]

As we have already seen in several examples, the electron withdrawing nature of the nitrogen atoms in heterocyclic systems leads to a marked increase in their ability to participate in cross-coupling reactions. The triazolo[1,5-*a*]quinazoline system in **8.23.** is a nice example, since not only the 5-chloro, but also its 5-tosyloxy derivative was shown to couple readily with 2-tributylstannyl-thiophene in 78% and 91% yield respectively.[32] Interestingly, in case of the chloro derivative the addition of copper improved the yield, while in the coupling of the tosylate it had an adverse effect.

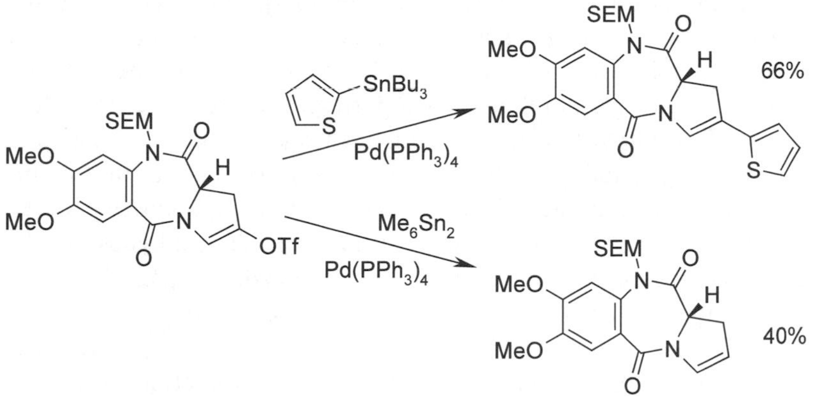

(8.22.)

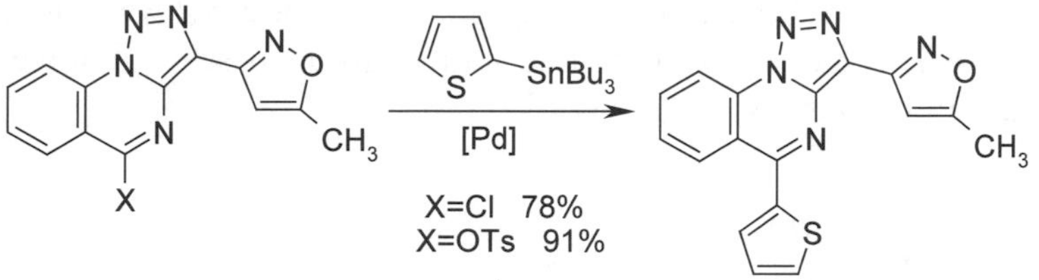

(8.23.)

The efficiency of different cross-coupling methods in the preparation of 2-imidazolyl-thieno[3,2-*b*]pyridine derivatives on the multihundred-gram scale was compared in a recent study. From the several methods tested (*e.g.* Negishi, Heck, Hiyama, Suzuki, Kumada, and Stille conditions) two stood out considerably. Negishi coupling of the imidazolylzinc reagent worked well up to 50g but was capricious upon scale up (**8.24.**). The only method that worked reliably on a >50g scale was the Stille coupling utilising a tributylstannyl-imidazole derivative, demonstrating the robustness of this method.[33]

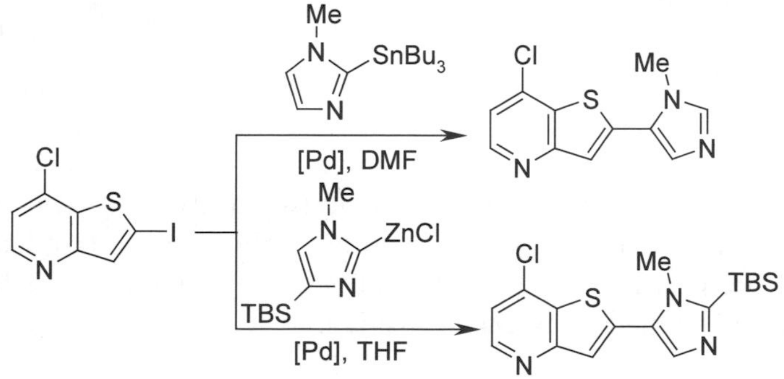

(8.24.)

Sonogashira coupling

The introduction of an ethynyl function onto the purine ring has been in the interest of pharmaceutical chemists for some time. Besides the Stille coupling of ethynyltin compounds, briefly mentioned in the previous chapter,[27] the Sonogashira coupling of halopurines and acetylene derivatives provides a straightforward access to this compound class. A series of nucleosides, bearing a halogen atom in the 8-position of the purine ring, were coupled with different acetylenes to increase their lipophilicity (**8.25.**). The range of acetylene derivatives span from long-chain alkynes such as octadecyne to sterane derived compounds and terpene derivatives.[34]

The preparation of ethynylpurines was achieved through the use of protected acetylenes. Hocek and co-workers coupled halopurines with trimethysilylacetylene to introduced the masked ethyne function, which was efficiently released on treatment with fluorides or base (**8.26.**).[35] The same compounds were also prepared by Hayashi and Kotschy, who used 2-methyl-3-butyn-2-ol as acetylene source and a strong base to remove the protecting group from the acetyl moiety.[36]

(8.25.)

(8.26.)

Although the Sonogashira-coupling of heterocycles is usually limited to their bromo and iodo derivatives, in certain cases the cross-coupling might also be achieved on activated chloro compounds. The chloro derivative shown in **8.23.** was coupled with trimethylsilylacetylene.[32] In another example the imidoyl chloride subunit of the 5-chloro-1,4-benzodiazepine derivative shown in **8.27.** coupled efficiently with phenylacetylene to give the expected disubstituted acetylene derivative.[15]

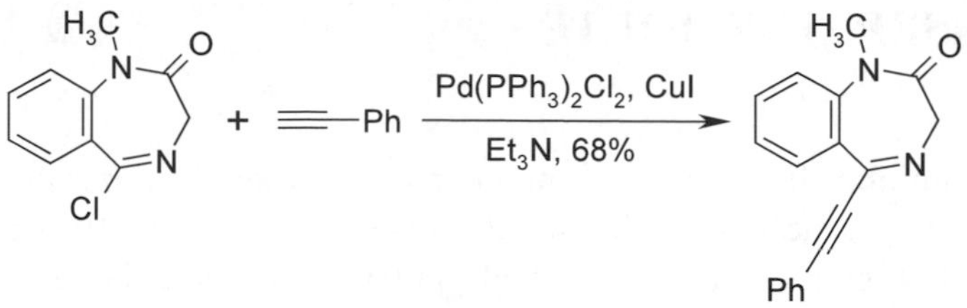

(8.27.)

The mildness of the Sonogashira coupling conditions allows for the functionalization of otherwise sensitive substrates too. The dibromo derivative of 2,1,3-benzothiadiazole was reacted with different pyridylacetylenes to give the expected fluorescent products (**8.28.**). The intermediate mono-coupled product was also isolated in each case, although in varying yield.[37]

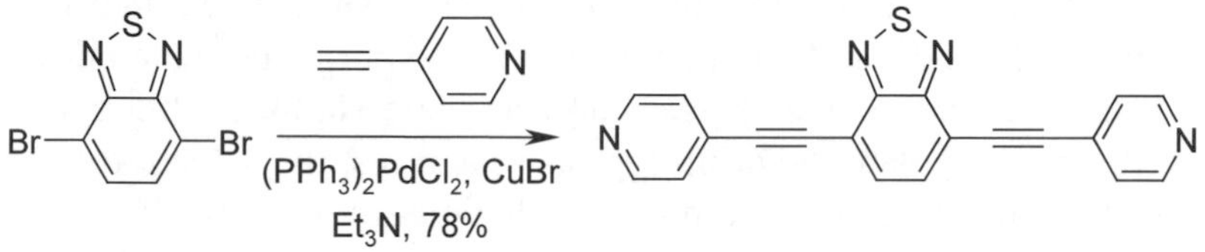

(8.28.)

The Sonogashira coupling was an effective tool for the functionalization of the 2-pyrone system, as reported by Fairlamb and co-workers. 4-Bromo-6-methyl-2-pyrone was reacted with a series of acetylene derivatives under both the Sonogashira and Negishi coupling conditions, using lithium acetylides in the latter case. Interestingly, the most effective catalyst system consisted of palladium on charcoal and triphenylphosphine (**8.29.**). The examined coupling methods usually gave similar results, although in certain cases their nature was complementary, only one or the other giving acceptable results.[38] The introduction of acetylene derivatives onto the analogous coumarin core has also been achieved recently using alkynytrifluoroborates in the presence of a palladium-dppf catalyst under neutral conditions.[39]

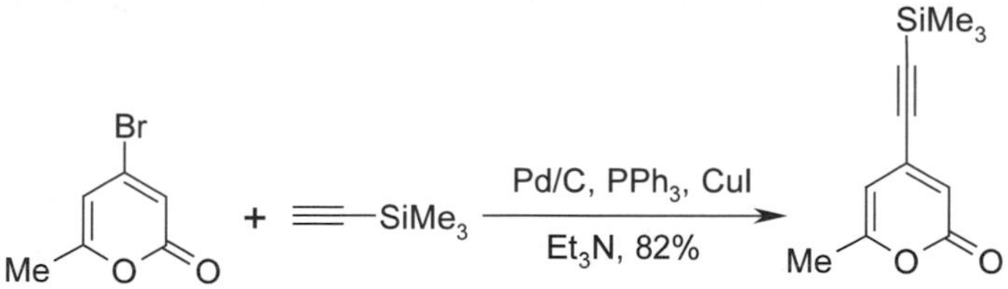

(8.29.)

8.2   INSERTION ROUTE

Unlike common five and six membered heterocycles, purines rarely undergo coupling reactions including the insertion of an olefin or carbon monoxide. This behaviour is not well understood since Heck and CO insertion reactions are known to proceed on similar systems.

Heck reaction (olefin insertion)

The preparation of ethenylpurines, due to the limited applicability of the Heck coupling on this skeleton, is usually achieved in a multistep fashion (see **8.20.**). including the introduction of the ethenyl functionality in a Stille reaction, followed by a Heck coupling on the established ethenylpurine. A unique example of the direct functionalization of the purine core includes the reaction of 8-bromocaffeine with *tert*-butyl acrylate (**8.30.**). Although the authors were able to isolate the desired product and test the selectivity of its binding to adenosine receptors, from the synthetic point of view the process is only of mediocre efficiency.[40]

$$H_2C-CH-CO_2{}^tBu \xrightarrow[\text{38\%}]{Pd(OAc)_2, (o\text{-tolyl})_3P}$$

(8.30.)

The Heck coupling of the purine analogue pyrazolo[1,5-a]pyrimidines on the other hand proceeds readily. A series of 3-iodopyrazolo[1,5-*a*]pyrimidine derivatives were reacted with different acrylates to give the expected coupling products as a single regioisomer in good to excellent yield (**8.31.**). The coupling was run in the presence of the common bis(triphenylphosphino)palladium dichloride catalyst using triethylamine as base. Interestingly the corresponding 3-bromopyrazolo[1,5-*a*]pyrimidines failed to undergo coupling under the same conditions.[41]

$$\xrightarrow[\text{91\%}]{\begin{array}{c}Pd(PPh_3)_2Cl_2\\ Et_3N,\ MeCN\end{array}}$$

(8.31.)

The oxazolo[4,5-*b*]pyridine system showed similar reactivity too. Its 6-bromo derivative underwent Heck coupling with methyl acrylate in DMF in the presence of the highly active palladium-tri-*o*-tolylphsophine catalyst system with good efficiency (**8.32.**).[42]

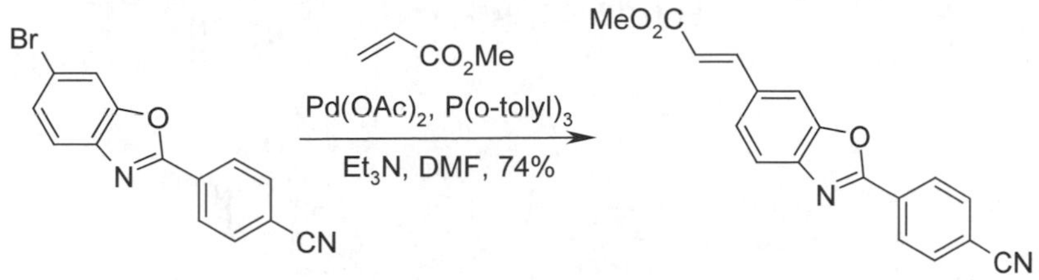

(8.32.)

Our survey of the relevant literature failed to produce a reaction where a heterocyclic ring system larger than 6 atoms participates in Heck coupling. A related example, which leads to a formally Heck coupled product, includes the reaction of *N*-vinyl-caprolactam with different vinyl-triflates (**8.33.**). The reaction was run both using conventional heating and microwave irradiation to give the desired product in good yield. Interestingly, the regioselectivity of the coupling seems slightly dependent on the way the heating is achieved.[43]

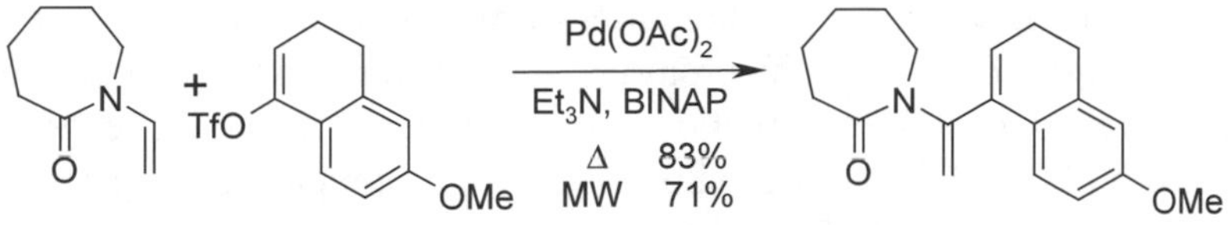

(8.33.)

CO-insertion (carbonylative coupling)

Although the insertion of carbon monoxide into hetarylpalladium complexes is well documented, the analogous reaction of purines is very rare.[44] Of the few examples of carbonylative coupling reactions on analogous systems two are presented here. The aza-derivative of 5-bromo-oxindole was reacted with different alcohols in the presence of a palladium catalyst and a slight overpressure of CO to give the corresponding carboxylic esters in poor to acceptable yield (**8.34.**).[45] The elevated temperature and pressure required for the coupling are characteristic of the carbonylation of azines.

In the other example CO insertion was used to introduced a ^{11}C-labelled amide function onto a series of aryl and hetaryl rings including oxindole and the β-carboline skeleton (**8.35.**). In the latter cast the trapping efficiency of [11]CO was 98% and the isolated yield of the labelled product was also good.[46]

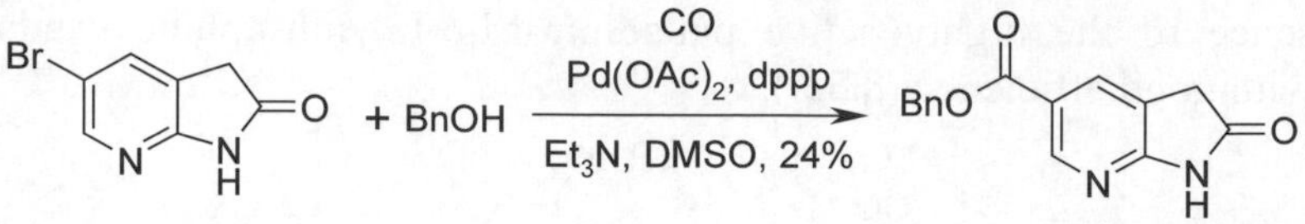

(8.34.)

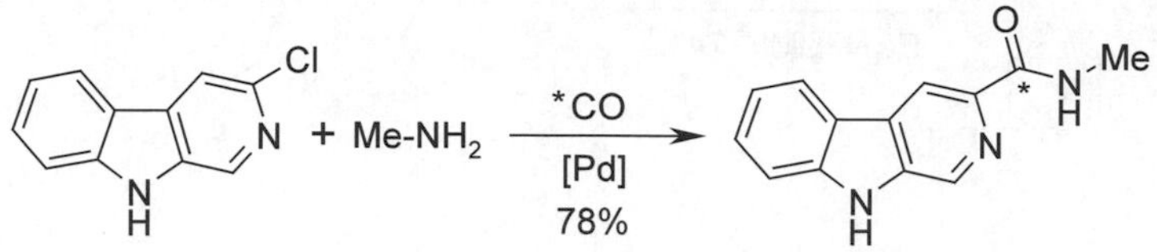

(8.35.)

8.3 CARBON-HETEROATOM BOND FORMATION

The exchange of a halogen to a classical nitrogen or oxygen nucleophile usually proceeds readily on the purine skeleton, without the necessity of using a transition metal catalyst. There are certain cases, however, where the palladium catalyzed carbon-heteroatom bond formation might take preference over noncatalysed methods. Inosine derivatives, for example, were reacted with ^{15}N-labelled benzamide in order to prepare the appropriate N-protected adenosines (**8.36.**). The coupling proceeded readily using a palladium-bis(diphenylphosphino)ferrocene catalyst system and caesium carbonate as base and after completion the removal of the acetyl protecting groups gave the desired adenosines. The method was successfully extended to the preparation of doubly labelled compounds too.[47]

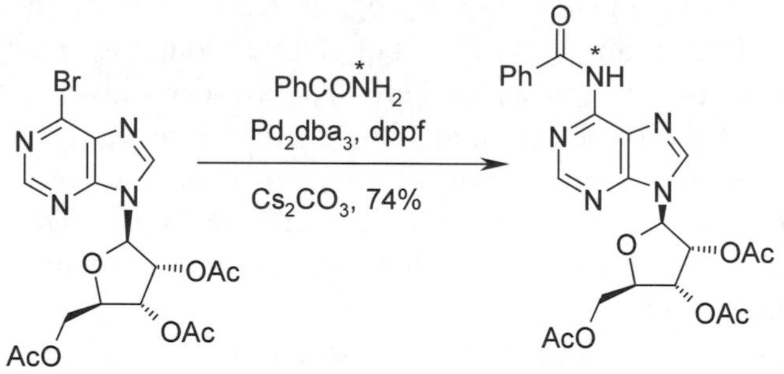

(8.36.)

The functionalization of the ring nitrogen atom in aziridines is traditionally achieved in classical substitution reactions. A recent article, however describes the transition metal catalyzed introduction of different aryl groups onto the aziridine core (**8.37.**). According to the reported results both the palladium catalyzed coupling of aryl halides and the copper catalyzed reaction of arylboronic acids proceed readily with cyclohexeneimine, although the latter procedure required considerably higher catalyst loadings.[48]

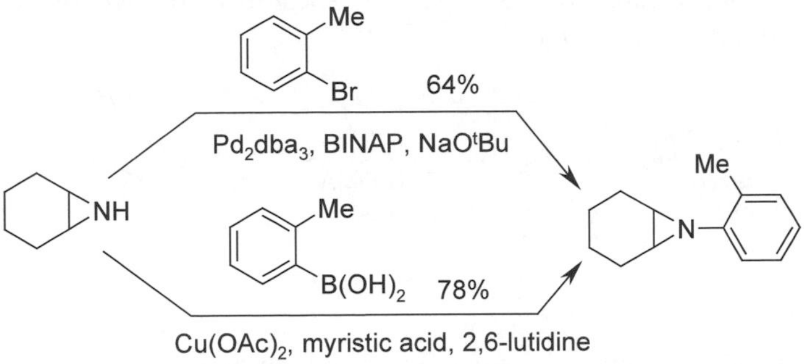

(8.37.)

The last example focuses not on the functionalization of heterocycles by a transition metal mediated carbon-heteroatom bond forming reaction, but the palladium catalyzed conversion of primary amines, including amino-heterocycles, into urea derivatives. A representative example, shown in **8.38.**, includes the reaction of an amino-carbazole derivative with morpholine, carbon monoxide and oxygen in the presence of catalytic amounts of palladium(II) iodide. The formation of the urea moiety proceeds with great selectivity and in high yield.[49] The reaction works equally well for primary aliphatic and aromatic amines.

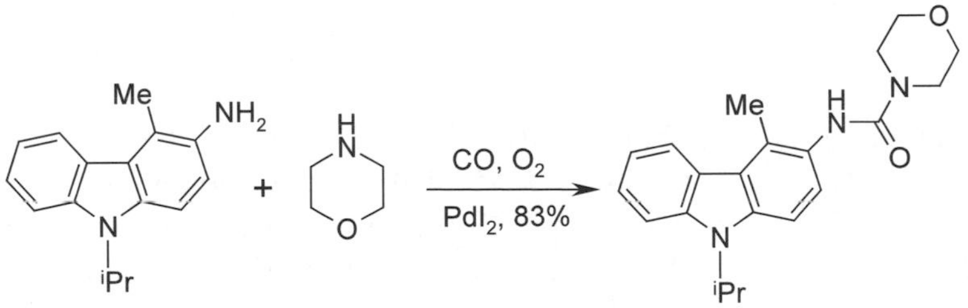

(8.38.)

8.4 OTHER PROCESSES

This chapter describes three different transition metal catalyzed reactions having one common feature. They are all examples of ring forming processes, leading to distinctly different product classes and following different mechanisms.

The copper mediated cycloisomerisation of alkynyl imines provides a facile access to pyrroles and fused aromatic pyrroloheterocycles.[50] In a recent example, shown in **8.39.**, the pyrrolopyrimidine bearing an acetylene moiety in the appropriate position cyclized smoothly to give the desired tricyclic compound in 89% yield.[51]

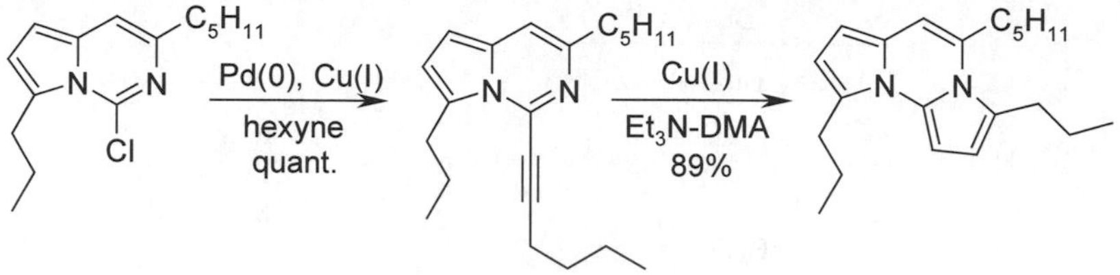

(8.39.)

The palladium catalyzed benzannulation reaction, described by Yamamoto, was successfully extended to the synthesis of condensed pyranone derivatives. The precursor to the cyclization underwent the benzannulation spontaneously under the applied Sonogashira coupling conditions (**8.40.**) on formation, to give the desired dibenzo[*b,d*]pyranone. The functionalities tolerated in the process include unsaturated bonds and polar functional groups, such as hydroxyl.[52]

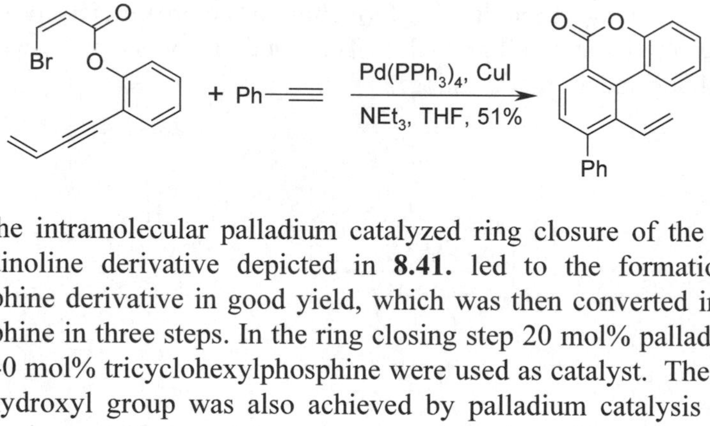

(8.40.)

The intramolecular palladium catalyzed ring closure of the tetrahydro-isoquinoline derivative depicted in **8.41.** led to the formation of the aporphine derivative in good yield, which was then converted into racemic aporphine in three steps. In the ring closing step 20 mol% palladium acetate and 40 mol% tricyclohexylphosphine were used as catalyst. The removal of the hydroxyl group was also achieved by palladium catalysis through its conversion to triflate and the subsequent reduction with ammonium formate in the presence of palladium acetate and dppf.[53]

(8.41.)

8.5   REFERENCES

[1] For recent reviews on the transition metal catalyzed functionalization of purines see: (a) Hocek, M. *Eur. J. Org. Chem.* **2003**, *245*. (b) Agrofoglio, L. A.; Gillaizeau, I.; Sato, Y. *Chem. Rev.* **2003**, *103*, 1875.

[2] (a) Miyaura, N.; Suzuki, A. *Chem. Rev.* **1995**, *95*, 2457. (b) Stanforth, S. S. *Tetrahedron* **1998**, *54*, 263. (c) Suzuki, A. *J. Organomet. Chem.* **1999**, *576*, 147.

[3] For a recent review on heterocyclic boronic acids see Tyrrel, E.; Brookes, P. *Synthesis*, **2003**, 469.

[4] Vollmann, K.; Müller, C. E. *Heterocycles* **2002**, *57*, 871.

[5] Fürstner, A.; Leitner, A.; Mendez, M.; Krause, H. *J. Am. Chem. Soc.* **2002**, *124*, 13856.

[6] Hocek, M.; Dvorakova, H. *J. Org. Chem.* **2003**, *68*, 5773.

[7] Havelkova, M.; Dvorak, D.; Hocek, M. *Synthesis* **2001**, 1704.

[8] Hocek, M.; Hockova, D.; Dvorakova, H. *Synthesis* **2004**, 889.

[9] Wu, T. Y. H.; Schultz, P. G.; Ding, S. *Org. Lett.* **2003**, *5*, 3587.

[10] Fraley, M. E. et al. *Bioorg. Med. Chem. Lett.* **2002**, *12*, 3537.

[11] Guillard, J.; Larraya, C.; Viaud-Massuard, M.-C. *Heterocycles* **2003**, *60*, 865.

[12] Lee, W. J.; Gee, M. B.; Yum, E. K. *Heterocycles* **2003**, *60*, 1821.

[13] Perron, J.; Joseph, B.; Mérour, J.-Y. *Tetrahedron* **2003**, *59*, 6659.

[14] Owens, A. P.; Nadin, A.; Talbot, A. C.; Clarke, E. E.; Harrison, T.; Lewis, H. D.; Reilly, M.; Wrigley, D. J.; Castro, J. L. *Bioorg. Med. Chem. Lett.* **2003**, *13*, 4143.

[15] Nadin, A.; Sanchez Lopes, J. M.; Owens, A. P.; Howells, D. M.; Talbot, A. C.; Harrison, T. *J. Org. Chem.* **2003**, *68*, 2844.

[16] Lee, J. H.; Park, E. S.; Yoon, C. M. *Tetrahedron Lett.* **2001**, *42*, 8311.

[17] Ryu, K.-M.; Gupta, A. K.; Han, J. W.; Oh, C. H.; Cho, C.-G. *Synlett* **2004**, 2197.

[18] Wu, J.; Wang, L.; Fathi, R.; Yang, Z. *Tetrahedon Lett.* **2002**, *43*, 4395.

[19] Fairlamb, I. J. S.; Syvänne, S. M.; Whitwood, A. C. *Synlett* **2003**, 1693.

[20] Bouisset, M.; Bousquet, A.; Heymes, A. DE3831533; *Chem. Abstr.* **1989**, *111*, 174139.

[21] Kotschy, A.; Nagy, A.; Bíró, A. B. PCT Int. Appl. (2004), WO 2004065386. *Chem. Abstr.* **2004**, *141*, 157130.

[22] Brathe, A.; Andresen, G.; Gundersen, L.-L.; Malterud, K. E.; Rise, F. *Bioorg. Med. Chem.* **2002**, *10*, 1581.

[23] Gundersen, L.-L.; Langli, G.; Rise, F. *Tetrahedron Lett.* **1995**, *36*, 1945.

[24] Hocek, M.; Votruba, I.; Dvorakova, H. *Tetrahedron* **2003**, *59*, 607.

[25] Shiota, T.; Yamamori, T. *J. Org. Chem.* **1999**, *64*, 453.

[26] Gundersen, L.-L.; Nissen-Meyer, J.; Spilsberg, B. *J. Med. Chem.* **2002**, *45*, 1383.

[27] Langli, G.; Gundersen, L.-L.; Rise, F. *Tetrahedron* **1996**, *52*, 5625.

[28] Brathe, A.; Gundersen, L.-L.; Nissen-Meyer, J.; Rise, F.; Spilsberg, B. *Bioorg. Med. Chem. Lett.* **2003**, *13*, 877.

[29] Havelkova, M.; Dvorak, D.; Hocek, M. *Tetrahedron* **2002**, *58*, 7431.

[30] Tiberghien, A. C.; Hagan, D.; Howard, P. W.; Thurston, D. E. *Bioorg. Med. Chem. Lett.* **2004**, *14*, 5041.

[31] For another illustrativ example of the functional group olerance of the Stille coupling see Kumamoto, H.; Tanaka, H. *J. Org. Chem.* **2002**, *67*, 3541.

[32] Jones, P.; Chambers, M. *Tetrahedron* **2002**, *58*, 9973.

[33] Ragan, J. A.; Raggon, J. W.; Hill, P. D.; Jones, B. P.; McDermott, R. E.; Munchhof, M. J.; Marx, M. A.; Casavant, J. M.; Cooper, B. A.; Doty, J. L.; Lu, Y. *Org. Proc. Res. Dev.* **2003**, *7*, 676.

[34] Flasche, W.; Cismas, C.; Herrmann, A.; Liebscher, J. *Synthesis* **2004**, 2335.

[35] Hocek, M.; Dvorakova, H.; Cisarova, I. *Collect. Czech. Chem. Commun.* **2002**, *67*, 1560.
[36] Hayashi, T.; Kawakami, T.; Kumazawa, H.; Nagy, A.; Csámpai, A.; Kotschy, A. PCT Int. Appl. WO 2003106458; *Chem. Abstrs.* **2004**, *140*, 42196.
[37] Akhtaruzzaman, Md.; Tomura, M.; Zaman, Md. D.; Nishida, J.-i.; Yamashita, Y. *J. Org. Chem.* **2002**, *67*, 7813.
[38] Fairlamb, I. J. S.; Lu, F. J.; Schmidt, J. P. *Synthesis* **2003**, 2564.
[39] Kabalka, G. W.; Dong, G.; Venkataiah, B. *Tetrahedron Lett.*, **2004**, *45*, 5139.
[40] Jacobson, K. A.; Shi, D.; Gallo-Rodriguez, C.; Manning, M. jr.; Müller, C.; Daly, J. W.; Neumeyer, J. L.; Kiriasis, L.; Pfleiderer, W. *J. Med. Chem.* **1993**, *36*, 2639.
[41] Yin, L.; Liebscher, J. *Synthesis* **2004**, 2329.
[42] Grumel, V.; Mérour, J.-Y., Guillaumet, G. *Heterocycles*, **2001**, *55*, 1329.
[43] Vallin, K. S. A.; Zhang, Q.; Larhed, M.; Curran, D. P.; Hallberg, A. J. Org. Chem. **2003**, 68, 6639.
[44] Tu, C.; Keane, C.; Eaton, B. *Nucleosides & Nucleotides* **1997**, *16*, 227.
[45] Cheung, M.; Hunter, R. N.; Peel, M. R.; Lackey, K. E. *Heterocycles*, **2001**, *55*, 1583.
[46] Karimi, F.; Langström, B. *Eur. J. Org. Chem.* **2003**, 2132.
[47] Terrazas, M.; Ariza, X.; Farràs, J.; Guisado-Yang, J. M.; Vilarrasa, J. *J. Org. Chem.* **2004**, *69*, 5473.
[48] Sasaki, M.; Dalili, S.; Yudin, A. K. *J. Org. Chem.* **2003**, *68*, 2045.
[49] Gabriele, B.; Salerno, G.; Mancuso, R.; Costa, M. *J. Org. Chem.* **2004**, *69*, 4741.
[50] Kel'in, J. T.; Sromek, A. W.; Gevorgyan, V. *J. Am. Chem. Soc.* **2001**, *123*, 2074.
[51] Kim, J. T.; Butt, J.; Gevorgyan, V. *J. Org. Chem.* **2004**, *69*, 5638.
[52] Kawasaki, T.; Yamamoto, Y. *J. Org. Chem.* **2002**, *67*, 5139.
[53] Cuny, G. D. *Tetrahedron Lett.* **2004**, *45*, 5167.

INDEX

Catalysis by Metal Complexes

Series Editors:
P.W.N.M. van Leeuwen, *University of Amsterdam, The Netherlands*
B.R. James, *University of British Colombia, Vancouver, Canada*

1. F.J. McQuillin: *Homogeneous Hydrogenation in Organic Chemistry.* 1976
    ISBN 90-277-0646-8

2. P.M. Henry: *Palladium Catalyzed Oxidation of Hydrocarbons.* 1980
    ISBN 90-277-0986-6

3. R.A. Sheldon: *Chemicals from Synthesis Gas.* Catalytic Reactions of CO and H_2.
    1983   ISBN 90-277-1489-4

4. W. Keim (ed.): *Catalysis in C_1 Chemistry.* 1983   ISBN 90-277-1527-0

5. A.E. Shilov: *Activation of Saturated Hydrocarbons by Transition Metal Complexes.*
    1984   ISBN 90-277-1628-5

6. F.R. Hartley: *Supported Metal Complexes.* A New Generation of Catalysts. 1985
    ISBN 90-277-1855-5

7. Y. Iwasawa (ed.): *Tailored Metal Catalysts.* 1986   ISBN 90-277-1866-0

8. R.S. Dickson: *Homogeneous Catalysis with Compounds of Rhodium and Iridium.*
    1985   ISBN 90-277-1880-6

9. G. Strukul (ed.): *Catalytic Oxidations with Hydrogen Peroxide as Oxidant.* 1993
    ISBN 0-7923-1771-8

10. A. Mortreux and F. Petit (eds.): *Industrial Applications of Homogeneous Catalaysis.*
    1988   ISBN 90-2772-2520-9

11. N. Farrell: *Transition Metal Complexes as Drugs and Chemotherapeutic Agents.*
    1989   ISBN 90-2772-2828-3

12. A.F. Noels, M. Graziani and A.J. Hubert (eds.): *Metal Promoted Selectivity in Organic
    Synthesis.* 1991   ISBN 0-7923-1184-1

13. L.I. Simándi (ed.): *Catalytic Activation of Dioxygen by Metal Complexes.* 1992
    ISBN 0-7923-1896-X

14. K. Kalyanasundaram and M. Grätzel (eds.): *Photosensitization and Photocatalysis
    Using Inorganic and Organometalic Compounds.* 1993   ISBN 0-7923-2261-4

15. P.A. Chaloner, M.A. Esteruelas, F. Joó and L.A. Oro: *Homogeneous Hydrogenation.*
    1994   ISBN 0-7923-2474-9

16. G. Braca (ed.): *Oxygenates by Homologation or CO Hydrogenation with Metal
    Complexes.* 1994   ISBN 0-7923-2628-8

17. F. Montanari and L. Casella (eds.): *Metalloporphyrins Catalyzed Oxidations.* 1994
    ISBN 0-7923-2657-1

Catalysis by Metal Complexes

* Volume 1 is previously published under the Series Title:
Homogeneous Catalysis in Organic and Inorganic Chemistry

euro design

– Frisidarie